AF759791

Integrated Disaster Risk Management

Editor-in-Chief

Norio Okada, School of Policy Studies, Kwansei Gakuin University,
Mita, Hyogo, Japan

Series Editors

Aniello Amendola, International Institute for Applied Systems Analysis (retired),
Laxenburg, Austria

Adam Rose, CREATE, University of Southern California, Los Angeles, USA

Ana Maria Cruz, Disaster Prevention Research Institute, Kyoto University,
Uji, Kyoto, Japan

Just the first one and one-half decades of this new century have witnessed a series of large-scale, unprecedented disasters in different regions of the globe, both natural and human-triggered, some quite new and others conventional. Unfortunately, this adds to the evidence of the urgent need to address such crises as time passes. It is now commonly accepted that disaster risk reduction (DRR) requires tackling the various factors that influence a society's vulnerability to disasters in an integrated and comprehensive way, and with due attention to the limited resources at our disposal. Thus, integrated disaster risk management (IDRiM) is essential. Success will require integration of disciplines, stakeholders, different levels of government, and of global, regional, national, local, and individual efforts. In any particular disaster-prone area, integration is also crucial in the long-enduring processes of managing risks and critical events before, during, and after disasters.

Although the need for integrated disaster risk management is widely recognized, there are still considerable gaps between theory and practice. Civil protection authorities; government agencies in charge of delineating economic, social, urban, or environmental policies; city planning, water and waste-disposal departments; health departments, and others often work independently and without consideration of the hazards in their own and adjacent territories or the risk to which they may be unintentionally subjecting their citizens. Typically, disaster and development tend to be in mutual conflict but should, and could, be creatively governed to harmonize both, thanks to technological innovation as well as the design of new institutions.

Thus, many questions on how to implement integrated disaster risk management in different contexts, across different hazards, and interrelated issues remain. Furthermore, the need to document and learn from successfully applied risk reduction initiatives, including the methodologies or processes used, the resources, the context, and other aspects are imperative to avoid duplication and the repetition of mistakes.

With a view to addressing the above concerns and issues, the International Society of Integrated Disaster Risk Management (IDRiM) was established in October 2009.

The main aim of the IDRiM Book Series is to promote knowledge transfer and dissemination of information on all aspects of IDRiM. This series will provide comprehensive coverage of topics and themes including dissemination of successful models for implementation of IDRiM and comparative case studies, innovative countermeasures for disaster risk reduction, and interdisciplinary research and education in real-world contexts in various geographic, climatic, political, cultural, and social systems.

More information about this series at http://www.springer.com/series/13465

Stefan Hochrainer-Stigler

Extreme and Systemic Risk Analysis

A Loss Distribution Approach

Springer

Stefan Hochrainer-Stigler
Risk and Resilience Program
International Institute for Applied Systems Analysis
Laxenburg, Austria

ISSN 2509-7091 ISSN 2509-7105 (electronic)
Integrated Disaster Risk Management
ISBN 978-981-15-2688-6 ISBN 978-981-15-2689-3 (eBook)
https://doi.org/10.1007/978-981-15-2689-3

© Springer Nature Singapore Pte Ltd. 2020
This work is subject to copyright. All rights are reserved by the Publisher, whether the whole or part of the material is concerned, specifically the rights of translation, reprinting, reuse of illustrations, recitation, broadcasting, reproduction on microfilms or in any other physical way, and transmission or information storage and retrieval, electronic adaptation, computer software, or by similar or dissimilar methodology now known or hereafter developed.
The use of general descriptive names, registered names, trademarks, service marks, etc. in this publication does not imply, even in the absence of a specific statement, that such names are exempt from the relevant protective laws and regulations and therefore free for general use.
The publisher, the authors and the editors are safe to assume that the advice and information in this book are believed to be true and accurate at the date of publication. Neither the publisher nor the authors or the editors give a warranty, expressed or implied, with respect to the material contained herein or for any errors or omissions that may have been made. The publisher remains neutral with regard to jurisdictional claims in published maps and institutional affiliations.

This Springer imprint is published by the registered company Springer Nature Singapore Pte Ltd.
The registered company address is: 152 Beach Road, #21-01/04 Gateway East, Singapore 189721, Singapore

Foreword to the IDRiM Book Series

In 2001, the International Institute for Applied Systems Analysis (IIASA) and the Disaster Prevention Research Institute (DPRI) joined hands in fostering a new, interdisciplinary area of integrated disaster risk management. That year, the two Institutes initiated the IIASA–DPRI Integrated Disaster Risk Management Forum Series, which continued over 8 years, helping to build a scholarly network that eventually evolving into the formation of the International Society for Integrated Disaster Risk Management (IDRiM Society) in 2009. The launching of the Society was promoted by many national and international organizations.

The volumes in the IDRiM Book Series are the continuation of a proud tradition of interdisciplinary research on integrated research management that emanates from many scholars and practitioners from around the world. This research area is still in a continuous process of exploration and advancement, several outcomes of which will be published in this Series. Efforts needed in the future to advance integrated disaster risk management include:

- Extending research perspectives and constructing new conceptual models
- Developing new methodologies
- Exploring yet uncovered and newly emerging phenomena and issues
- Testing important hypotheses
- Engaging in proactive field studies in regions that face high disaster risks, and performing such studies that incorporate research advances in disaster-stricken regions
- Designing new policies to implement effective integrated disaster risk management

Obviously, the above approaches are rather interdependent, and thus integrated disaster risk management is best promoted by combining them. For instance, emerging mega-disasters, which are caused by an extraordinary natural hazard taking place in highly interconnected societies may require a combination of both the second and third points above, such as disaster governance based in part on mathematical models of systemic risks. Also, long-range planning for societal implementation of integrated disaster risk management inevitably requires encompassing most of the above approaches.

The IDRiM Book Series as a whole intends to cover most of the aforementioned new research challenges.

Nishinomiya, Japan	Norio Okada
Milan, Italy	Aniello Amendola
Laxenburg, Austria	Joanne Bayer
Uji, Japan	Ana Maria Cruz
Laxenburg, Austria	Stefan Hochrainer-Stigler
Los Angeles, CA, USA	Adam Rose
Boulder, CO, USA	Kathleen Tierney
Oberlin, OH, USA	Ben Wisner

Contents

List of Figures

List of Tables

Chapter 1
Introduction

This book is about how extreme and systemic risk can be analyzed in an integrated way. Risk analysis is understood here to include measurement, assessment as well as management aspects. Integration is understood as being able to perform risk analysis for extreme and systemic events simultaneously. In this introductory chapter, we want to provide some motivation for such an integration and also want to give a short overview of the topics presented in this book. In doing so it is useful to first discuss the concepts, approaches used as well as their importance in some detail for both types of risks separately. The definition of extreme and systemic risk plays a pivotal role in that regard and therefore provides the starting point for our discussion.

1.1 Risk Analysis of Extremes: Loss Distributions

There are various ways how to define what an extreme event is. For example, the Emergency Events Database (EM-DAT), the worlds most comprehensive data on the occurrence and effects of disasters from 1900 to the present day, of the Centre for Research on the Epidemiology of Disasters (CRED) uses a threshold approach. Only if at least one of the following four criteria for a disaster are fulfilled it is entered into the database: Ten (10) or more people reported killed, Hundred (100) or more people reported affected, Declaration of a state of emergency, or a Call for international assistance (EM-DAT 2019). Using such a definition to indicate a disaster or extreme event one can gather information about their occurrences over given time periods as well as their importance. For example, one can find that between 1998 and 2017, more than 7,200 events were recorded in EM-DAT, 91% of which were climate-related. Furthermore, one is able to find out that during this 20-year time period, floods were the most frequent type of disasters, comprising about 43% of total events. They also affected the largest number of people (more than 2.0 billion), followed by drought (around 1.5 billion people) (Wallemacq and House 2018). These numbers already indicate that extreme events present a serious threat

© Springer Nature Singapore Pte Ltd. 2020

S. Hochrainer-Stigler, *Extreme and Systemic Risk Analysis*, Integrated Disaster Risk Management, https://doi.org/10.1007/978-981-15-2689-3_1

around the world. Indeed, recent global calls such as the Sendai Framework for Risk Reduction (United Nations 2015b) or the Sustainable Development Goals (United Nations 2015a) place special emphasis on such risks and possible ways to reduce them. Also in light of future climate change impacts the database indicates that increases in losses must be expected, as most extreme events are weather-related (SREX 2012). Hence, extreme events already deserve special attention in its own right (Collins et al. 2017).

However, other databases such as the one by Munich Reinsurance (one of the largest reinsurers for catastrophe risk in the world) apply different thresholds with an emphasis on normalized economic losses (in 2016 USD) and in respect to income country groups. For example, an extreme event to be categorized in Cat Class 3 within the Munich Re NatCat database is informally defined to be an event with a large loss having a major impact and formally defined as an event which causes losses above 300 million USD in a high-income country group (and with more than 100 fatalities). The loss threshold for low-income country groups is considerably lower in that class with a value of only 10 million USD which needs to be exceeded. Hence, Munich Re takes not only losses into account but also to some extent the degree of severity to a country (Munich Re 2018). Using this database one can find that in low-income countries, the percentage of insured losses from natural disasters is negligible, highlighting a distributional aspect to past disaster losses. This is an important point to be considered. For example, from 1998 to 2017, total reported losses from climate-related disasters amounted to around 1,432 billion USD for high-income countries and 21 billion USD for low-income countries (Wallemacq and House 2018). However, while economic losses were considerably higher for high-income countries, the economic burden was in fact much greater for low-income countries, e.g., average losses in terms of the Gross Domestic Product (GDP) were only around 0.4% in high-income countries but 1.8% in low-income countries. Similar patterns can be found also if one considers the number of deaths or the number of people affected by disasters (for a detailed analysis see Wallemacq and House 2018, Munich Re 2018). As this discussion already indicates many other possible threshold levels as well as approaches, how to define extreme events or disasters can be found in the literature (SREX 2012; United Nations 2016). As suggested in this book, a large part of the confusion about how an appropriate threshold should be chosen to define extremes can be avoided if they are treated within the context of random variables as discussed next.

To move forward we already now need to introduce some notations and definitions. We keep them in this introductory chapter to a minimum and will refer to the details to the other chapters of this book. A random variable X is a real-valued function defined on a probability space and is characterized by its cumulative distribution function F with $F(x) = P(X \leq x)$. In other words, the cumulative distribution function (or simply distribution) of a random variable X is the probability that X is less than or equal to a given number x, notated as $P(X \leq x)$. For example, tossing a coin can be modeled as a random variable using the Binomial distribution, rainfall can be modeled as a random variable using a Gamma distribution. Indeed, many different types of distributions exist, most prominently the Normal distribution, that can be

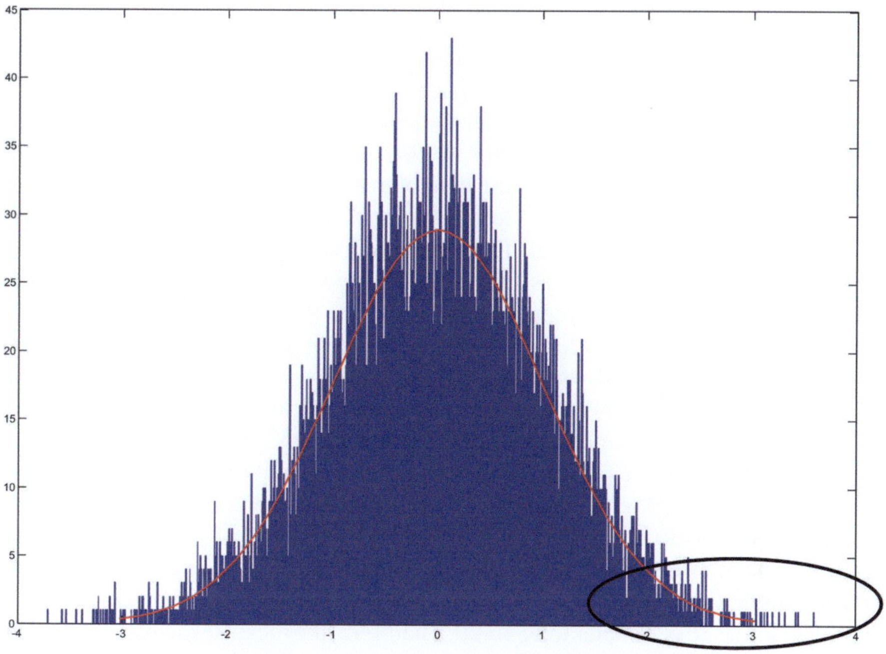

Fig. 1.1 Histogram of random samples (sample size = 1000) from a standard normal distribution $N(0, 1)$ and fitted curve

selected as a possible candidate to specify the random variable X. The standard approach is that one tries to fit a distribution to the entire dataset of available past observations, e.g., losses, using maximum-likelihood techniques and afterward uses goodness of fit statistics to determine its appropriateness (Sachs 2012). However, while a distribution model may be chosen for its overall good fit to all observations, it may not provide a particularly good fit for extreme events. Indeed, the distributional properties of extremes are determined by the upper and lower tails of the underlying distribution. Unfortunately, while standard estimation techniques fit well where data has greatest density, they can be severely biased in estimating tail probabilities (see Fig. 1.1). Hence, a theory of its own is needed for extremes, so-called Extreme Value Theory (EVT) (Embrechts et al. 2013).

The starting point to characterize the behavior of extremes within EVT is by considering the behavior of the sample maxima $M_n = \max\{X_1, X_2, ..., X_n\}$, where $X_1, ..., X_n$ is a sequence of independent random variables having a common distribution function F (Embrechts et al. 2013). Surprisingly, the distribution of extremes is largely independent of the distribution from which the data is drawn. Specifically, the maximum of a series of independent and identically distributed (i.i.d.) observations can take only one of three limiting forms (if at all), the Fréchet, the Weibull or the Gumbel distribution (McNeil et al. 2015). However, one major challenge in estimating accurately the tail of the distribution is data scarcity. Indeed, most data is (naturally) concentrated toward the center of the distribution and so, by definition, extreme data is scarce and therefore estimation is difficult (see the circle in Fig. 1.1). Two important kinds of modeling approaches for estimating extreme events were developed in the past to account for this situation: the (traditional) *block maxima* models and (more modern) models for *threshold exceedances* (Coles et al. 2001). The block maxima approach is looking at maximas over blocks, e.g., the largest loss in a given year, while the threshold approach looks at maximas defined to be above a (pre-determined) threshold level. The distribution of maximas and distributions of exceedances are providing the basis for modeling and estimating the tail behavior of real-world data, including extreme events.

For illustration, we take a look at rainfall data from a weather station in Malawi (in Chitedze) where rainfall was measured from 1961 till 2005 on a daily basis (see Hochrainer et al. 2008). This data will be also used extensively in other chapters as well. Rainfall is often modeled through a Gamma distribution and we therefore select this distribution as our candidate. We focus here on dekad rainfall (total amount of rainfall over 10 day periods) in the rainy season (selected here to be from October to March, see Nicholson et al. 2014). After applying a method of moments technique (see Sect. 2.1) to estimate corresponding parameters of dekad rainfall for this period and performing some goodness of fit tests, we eventually find that the shape parameter is 0.99 and the scale parameter to be 46.00. Therefore, the mean dekad rainfall amount is around 45.75 mm. The probability that rainfall is smaller or equal than 150 mm is calculated to be 96%, or in other words with only 4% probability dekad rainfall in the rainfall season is above 150 mm. This indicates that such a situation only occurs very rarely. Some additional analysis gives further indications for us that especially the tails are not fitted very well. We therefore apply a block maxima approach and therefore only take the maximum dekad rainfall for each year and season into account. We first recognize that we now only have 44 data points available and we use EVT to find that a Gumbel distribution is appropriate here, e.g., using diagnostic plots. The related location and scale parameters are estimated to be 127.57 and 30.59. Assuming a correct model with the estimates given above one can now investigate in more detail extreme events. For example, the probability that the maximum dekad rainfall in the rainy season is below 150 mm can be calculated as (using the definition of a Gumbel distribution, see Sect. 2.2) $P(X \leq 150) = \exp\left(-\exp\left(-\frac{150-127.57}{30.59}\right)\right) = 0.62$. In other words, given our Gumbel distribution assumption, the probability that the maximum dekad rainfall over the rainy season will be greater than 150 is 38% or will occur, on average, every 2.6 (=1/0.38) years. A quite different result than one would expect without an EVT analysis. One could now build a mixed distribution by taking both distributions, the Gamma as well as the Gumbel distribution (or in other words one for the non-extreme cases and one for the extreme cases) explicitly into account (Miller 2018). In this book, we again call them simply the distribution and if the random variable represents pure downside risk, e.g., losses, we will call them a *loss distribution*.

Given that one is able to estimate a distribution as discussed above, there is the question what to do with it. A distribution is actually a complex object and there is a need to describe it with relevant parameters. The actual decision which measures should be used very much depends on what type of risks a decision-maker is interested in. Frequently used location measures include the *Expectation*, the *Median*, and *Mode*. The most prominent dispersion measures include the *Variance*, the *Standard Deviation* as well as the *Mean Absolute Deviation* (Pflug and Roemisch 2007). The expectation, median, and mode are used in many applications that are especially interested in the average behavior of a random variable and the variance and standard deviation are the primary risk measures for the behavior (e.g., fluctuation) around the mean (Mechler 2016). All of them more or less focus on the area of the distribution with the highest density (see Fig. 1.1). However, in recent times there has been a growing interest in measures for the tails of a distribution, including measures such as the *Value at Risk* or the *Expected Shortfall* (Abadie et al. 2017). Importantly, evaluations of these measures have the potential to inform risk management strategies aiming to decrease extreme risk.

As suggested in this book a so-called risk-layer approach is very helpful to identify appropriate options to decrease risk using a loss distribution as the starting point (Benson et al. 2012). In the case of natural disasters, risk layering involves identifying interventions based on hazard recurrences (see Fig. 1.2). The approach is based on the assumption that varying risk management strategies adopted by different risk bearers (households, businesses, or the public sector) are appropriate for distinct levels of risk on the grounds of cost-efficiency and the availability of risk

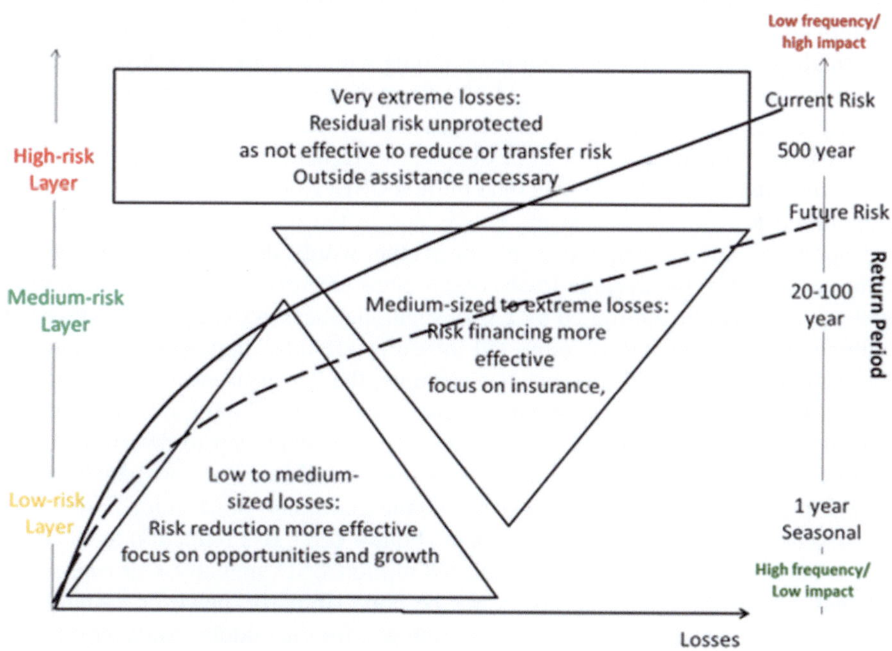

Fig. 1.2 Risk-layer approach for risk management using a loss distribution. *Source* Adapted based on Hochrainer-Stigler et al. (2018a)

reduction and financing instruments (Mechler et al. 2014). Hence, risk measures can be related to the most appropriate instruments to decrease risk. For example, for low- to medium-loss events that happen relatively frequently, risk reduction can be assumed to be cost-effective in reducing burdens. Such events are usually assessed using the expectation of reduced losses to estimate the effectiveness of risk reduction measures, for example, via cost-benefit analysis (Mechler 2016). As the costs of risk reduction often increase disproportionately with the severity of the impacts, other instruments need to be considered as well (Hochrainer 2006). As discussed in Linnerooth-Bayer and Hochrainer-Stigler (2015) it is generally advisable to use risk-financing instruments mainly for lower probability/higher impact extremes that have debilitating consequences; these are usually measured through deviation and tail measures. Finally, there is an uppermost layer, where it is generally too costly to use even risk-financing instruments against very extreme risks, and global pooling may be needed.

For illustration purposes, we use estimated loss distributions for Afghanistan taken from the Aqueduct global flood analyzer (Winsemius et al. 2016) which among only a few ones provides information on urban flood risk on the country level. In more detail, losses for nine different return periods are available there. For example, a 5-year event loss, e.g., a loss that happens on average every 5 years, would cause losses of about 65 million USD while a 100-year event would cause losses of about 229 million losses. If we assume that the low risk layer includes all events between the 1- and 100-year return period, the middle risk layer all events that happen with a return period between the 100- and 500-year event, and all events which happen less frequent than a 500-year event period to be within the high risk layer, one can determine the average annual losses (AAL) for each risk layer (through integration of the respective parts of the loss distribution, see Sect. 2.1), which can in turn indicate costs to reduce risk in each layer. For example, for Afghanistan AAL for the three risk layers are found to be 35, 0.38, and 0.05 million USD. Note, the losses are especially pronounced for the low risk layer. This is due to the fact that averages are more appropriate to indicate frequent events, or in other words, due to the low probability nature of extremes the average losses of extremes are very small even if losses are large (as they will be multiplied by a very small number). As suggested, to take the extreme events better into account, tail measures are more appropriate to indicate the magnitude of a possible event. For example, the 250-year event loss is around 270 million USD and the 1,000-year loss event is around 337 million USD. The risk estimates for each risk layer can be used in a subsequent step to determine costs for risk reduction. For example, if one uses overall estimates of cost-benefit ratios for risk reduction to be around 1:4 (see also the discussion in Mechler 2016), e.g., investment of 1 USD will reduce losses of about 4 USD, risk reduction to decrease risk in the first layer would cost around 8.8 million USD annually. Corresponding calculations for the other risk layers can be done similarly; however, usually tail measures are more appropriate here. For example, for the middle risk layer one can use the actuarial fair premium and multiply it by 20 (Hochrainer 2006) as well as add a corresponding tail measure index (e.g., a 250-year event loss) to determine the costs for risk financing (see, for example, Jongman et al. 2014), which for Afghanistan would result in annual costs of about 278 million USD. This may be seen as very expensive but in light of possible reduction of additional indirect costs, e.g., due to business interruption, could still be considered as one feasible option in the future. It should be noted that a risk layer approach is especially appropriate in cases where risk can be quantified and less adequate if other dimensions, such as loss of life, should be considered as well. The different approaches and assumptions which can be used as well as advantages and disadvantages of different risk management options will be discussed in detail in Chaps. 2 and 3. Summarizing, a loss distribution approach can be seen as a promising way forward to measure, model, and manage extreme event risk and a risk layer approach is useful to determine risk management strategies to decrease or finance such risks. This ends our short discussion on extreme risk analysis and a comprehensive discussion will be given in Chap. 2 of this book. We now move forward to discuss systemic risk in some detail.

1.2 Systemic Risk Analysis: Dependencies and Copulas

As in the case of extreme risk before the term systemic risk can be introduced, it is necessary to establish an understanding of what one means under the term *system*, what kind of systems exist (or may be assumed to exist, see Luhmann et al. 2013), and how systems can be analyzed. We use the main focal point in systemic risk analysis, namely, the emphasis on the connection between elements within a system (Helbing 2013), to start our discussion. In doing so we first note that a system is usually comprised of individual elements. For example, if a system consists of only one individual element and is "at risk" it could be treated as individual risk with the methods as discussed in the section before (e.g., estimating a loss distribution). Even in the case of many unconnected individuals which are "at risk" in the system, one could treat each element as carrying individual risks only and again one could apply the techniques as discussed in the previous section. The situation changes if one assumes some kind of interaction between the elements of a system. If an element in a system is (somehow) dependent or influenced by another element in the system an additional aspect must be introduced, namely, an interaction or dependency dimension. Interactions between elements can eventually produce higher order effects, commonly subsumed under the term emergence (Barabási and Albert 1999), and such higher order systems can behave quite differently than the elements they consist of (Luhmann 1995). Conceptually, everything could be seen as either an element of a system or a system itself consisting of elements (see, for example, Elias 1987). As Fig. 1.3 indicates various different stages of embedded systems can be looked at from this perspective. It is obvious that the complexity of the systems in Fig. 1.3 increases from the bottom to the top. One possibility to define this complexity is in terms of the "structuredness" of a system; the properties of the composite elements are increasingly more dependent on the organization of the elements and less dependent on the actual properties of the isolated elements (for a comprehensive discussion see Thurner et al. 2018). Consequently from a system's perspective at

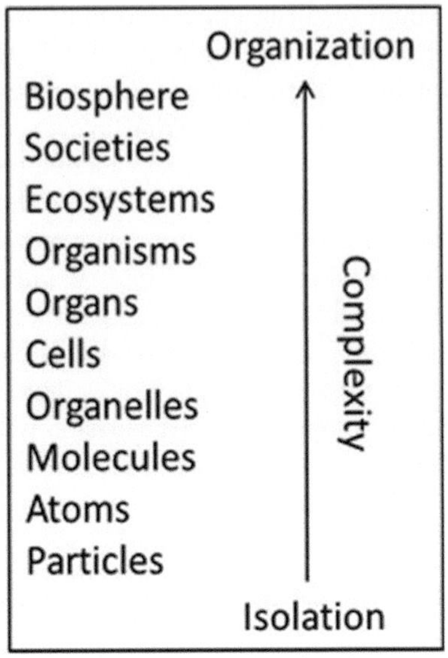

Fig. 1.3 A simple hierarchy of integrated systems, complexity, and unit of analysis

some point, the focus on isolated observed characteristics is less adequate than the focus on the organization between the elements. Methodological wise this implies that the modeling of interdependent relations and processes becomes increasingly necessary, and this is especially true for systemic risk analysis.

The distinguishing feature of systemic risk analysis is its emphasis on the connection between individual elements within a system. Therefore, systemic risk is also sometimes called network risk (Helbing 2013). Realization of systemic risk leads, by definition, to a breakdown or at least major dysfunction of the whole system (Kovacevic and Pflug 2015). Systemic risk can realize exogenously, e.g., through outside attacks, or endogenously, e.g., due to sudden failures in network nodes. Dependent on the system and its properties various different situations can emerge leading sometimes to systemic risk as elements in the system are too big to fail, or too interconnected to fail or so-called keystone elements which in times of failure cause large secondary effects or to a complete breakdown of the system (see Hochrainer-Stigler et al. 2019 for a comprehensive discussion). More generally, systemic risk is usually due to complex cascading effects among interconnected elements ultimately leading to collective losses, dysfunctions, or collapses. A lot of current research on systemic risk is done within the financial domain, especially for banking systems. There, the emphasis is usually on the interdependencies and contributions of individual elements to systemic risk. However, a re-focus on systemic risk dimensions in many other research domains can be currently observed. Part of this phenomenon can be explained by the ever increasing complexity of the world as well as data availability, e.g., due to digitalization (Helbing 2018).

How is systemic risk treated in the various scientific disciplines? While a comprehensive discussion will be given in later chapters of this book (see also Hochrainer-Stigler et al. 2019), some indicative examples are worthwhile to be noted already here. For example, in the ecological domain, experiments were run in the past to measure impacts of perturbation in controlled ecosystems, including secondary species extinction or total biomass reduction. Different types of shocks such as pulse or press perturbations or just the simple removal of species and related consequences were studied there (Scheffer and Carpenter 2003). In the socio-economic domain, various case studies were performed in the past to identify mechanisms that can cause the realization of systemic risks in societies (Frank et al. 2014) or specific sectors, such as the banking sector (Battiston et al. 2012). The identification of the mechanisms and corresponding dynamics that cause a system to collapse can give important indications of how to decrease systemic risk. For example, Folke et al. (2004) did a comprehensive review on the evidence on regime shifts in terrestrial and aquatic ecosystems and the authors noted that regime shifts can occur more easily if resilience has been reduced, e.g., due to the removal of specific groups. As will be discussed in later chapters, the concept of resilience as defined by Holling (1973) can provide a promising entry point for the integration of extreme and systemic risk management strategies.

As the examples above should point to, systemic risk (similar to extreme risk) has a wide area of application and will get even more important in our increasingly complex (and therefore interdependent) world (Centeno et al. 2015). The diverse applications also indicate that many different phenomena on how systemic risk can realize are possible. A very beneficial perspective on systemic risk for our purpose is by treating a system as a network. A system will be defined as a set of individual elements (e.g., the nodes of the network) which are (at least partly) connected or dependent with each other (see Fig. 1.4). It can already be noted here that the nodes in such a network perspective are in our case random variables with a corresponding loss distribution. We will call the elements (or nodes, or sometimes also called agents) which are "at risk" in the system as *individual risk* throughout the book.

From a broad-based perspective, two quite distinct starting points for analyzing systemic risk within a network approach can be distinguished. Either one starts with a network and some characterizations of it including the topology, the strength between nodes, loadings, possible multi-layered structures, position of leaders and followers, overlapping portfolios, possible regulations, initial conditions, and populations dynamics (to name but a few) *or* one defines the relevant elements and interaction behaviors and let the network build up by itself over time. In the later

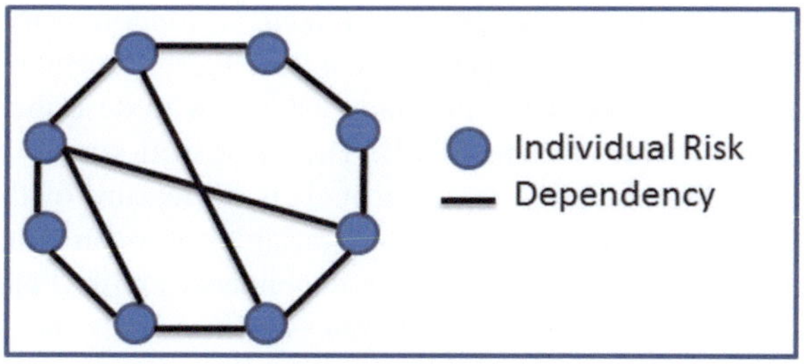

Fig. 1.4 Example of a system as a network comprised of individual risks and dependencies

case, this includes so-called agent-based modeling approaches but can also be done in other ways, for example, using differential equations as a starting point (Heppenstall et al. 2011). The former cases are usually used for theoretical analysis but most often also include some (rather simple) empirical examples to strengthen the messages derived analytically or via simulation. They are usually not as data intensive as in the case of agent-based modeling approaches. There, one relies on specific high detailed micro-data to calibrate the model to reflect past macroeconomic behavior. Hence, agent-based models are usually high resource-intensive tasks and also have various drawbacks in regards to the interrelation of micro- and macro-level behavior, but on the other hand are able to model emergence. Nevertheless, for both approaches at some point one ends up with a network which is either static (e.g., going to one equilibrium) or dynamic (e.g., with multiple equilibria or just patterns) and, generally speaking, usually the system has some stable regions as well as ones with high systemic risk. At this point, various measures for systemic risk can be applied to analyze the behavior of the system.

As in the case of extreme risk also for systemic risk, many measures have recently been suggested, particularly for financial related systemic risks (due to their importance for society, as well as the availability of high-precision and high-resolution data). One of the most important systemic risk measures in financial network analyses today is the so-called *DebtRank* (Battiston et al. 2012). DebtRank estimates the impact of one node, or group of nodes, on the others and is inspired by the notion of centrality in a network. Accordingly, DebtRank can be considered as a warning indicator for a node being too central to fail; an important feature aggravating a node's contribution to systemic risk. However, other measures of systemic risk are also available, such as the Systemic Expected Shortfall (SES), which uses thresholds to quantify a node's anticipated contribution to a systemic crisis. Based on these systemic risk measures, strategies to decrease risk can be developed. For example, Poledna and Thurner (2016) proposed a risk measure based on DebtRank that quantifies the marginal contribution of individual liabilities in financial networks to the overall systemic risk. The authors use this measure to introduce an incentive for reducing systemic risks through taxes, which they show can lead to the full elimination of systemic risks in the systems considered. The resultant proposal of a systemic risk tax can be seen as a concrete measure that can increase individual and systemic resilience if implemented.

From our suggested network perspective of a system, the aforementioned interconnections between the elements will be interpreted as dependencies and as our elements in the system are treated as random variables, one needs appropriate tools on how to measure and model these dependencies (Fig. 1.4). It is common knowledge that there are many ways how to describe the dependence between random variables (Sachs 2012). The most well-known measure of dependence is the Pearson correlation coefficient. It is a measure of the strength of the linear relationship between two random variables and can take values between -1 (inverse dependence) and $+1$ (positive dependence) with 0 indicating no association. However, this correlation coefficient is a linear measure and therefore the dependence is assumed to have the same strength across the respective range of the random variables considered. It therefore cannot answer questions like, if one risk has a very large loss, e.g., an extreme event has happened, how more likely it is that another risk also has a large loss, e.g., that another element in the system also experiences a large loss or an extreme event. There are situations that exhibiting so-called tail dependence, i.e., stronger dependencies between random variables in the tails of the respective distributions. This phenomenon is, for example, noticeable in hydrological processes such as water discharge levels across basins. For example, in the work of Jongman et al. (2014), it was shown that large-scale atmospheric processes can result in strongly correlated extreme discharges across river basins in Europe.

To address the issue of tail dependence, so-called copula approaches are superior compared to other dependence measures. For example, a copula can *explicitly* address the question if one risk has a very large loss, how much more likely it is that another risk also has a large loss. In Fig. 1.5 random samples from a specific type of copula (i.e., the flipped Clayton copula, see Chap. 3) which was estimated based on empirical discharge data for three different basins in Romania are shown. The figure indicates that while during frequent events (e.g., small values on the x- and y-axes in each of the pair-wise plots) there are large fluctuations in the individual river discharges, for extreme events (e.g., for large values on the x- and y-axes) there are less fluctuations. In other words, extreme discharge levels in one basin are usually accompanied with

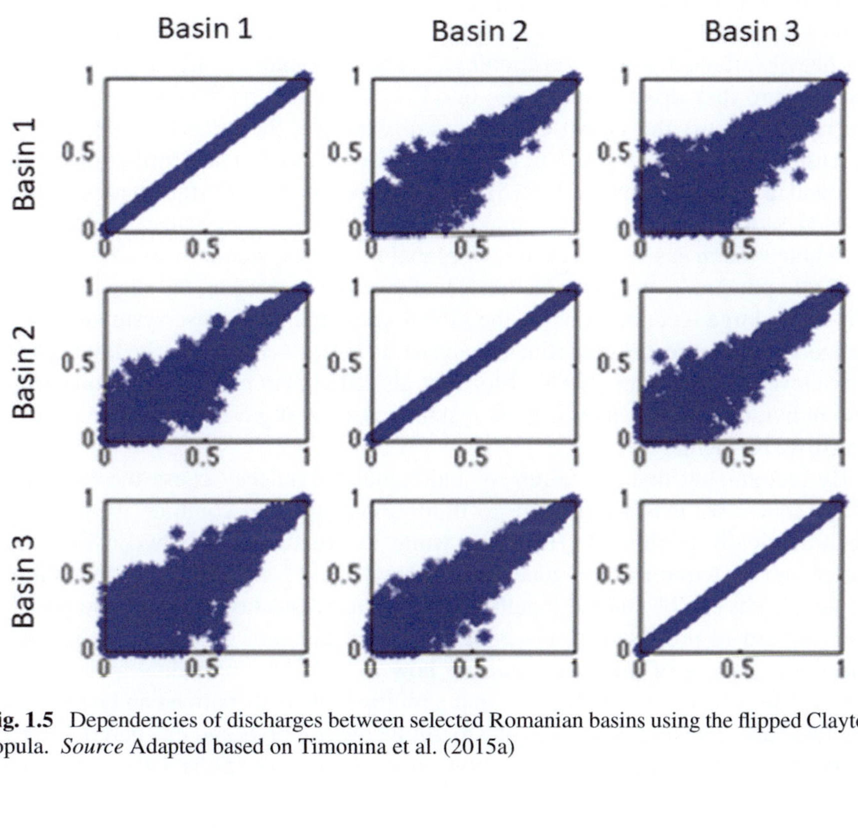

Fig. 1.5 Dependencies of discharges between selected Romanian basins using the flipped Clayton copula. *Source* Adapted based on Timonina et al. (2015a)

extreme discharge levels in other basins. One important consequence is the fact that this can lead to large-scale flooding with severe consequences including systemic risk realization (Timonina et al. 2015b). This ends our short overview of systemic risk measures and approaches currently applied. A more detailed discussion will be given in Chap. 3. We proceed now to the main question addressed in this book on how both individual and systemic risk analysis can be integrated.

1.3 Extreme and Systemic Risk Analysis: Sklar's Theorem

As in the previous two sections also here it is worthwhile to emphasize the relevancy of an integrated approach with some examples first. Past experiences have shown the potential of extreme events to trigger systemic risks (see also Little 2002). As a case in point, the 2011 Thailand floods and its worldwide consequences have demonstrated the magnitude of potential knock-on effects. Indeed, the prolonged floods had a devastating impact on both the world and the Thailand economy. For example, a decrease in real GDP growth rate in 2011 from an expected 4.1 to 2.9% has been estimated. Furthermore, a decrease in world's industrial production by 2.5% due to this event was projected. Especially the automobile sector in Japan suffered huge losses primarily because some critical components could not be delivered as one major component producer was inundated. Additionally, also the electronic sector was heavily affected. Before the flooding, Thailand produced nearly half (43%) of the world's hard disk drives. Due to the flood event production declined considerably and rippled across the global economy. For example, the price for desktop HDD increased by around 80–190% (a comprehensive discussion and further details can be found in Haraguchi and Lall 2015). What this points out is the fact that the world is nowadays closely interconnected making it possible that events can cascade through the whole system easily. In a review of global relevant systemic risks Centeno et al. 2015 also put strong emphasis on individual events that, even on the very local level, may cause large repercussions on the global scale and may cause systemic risks to realize. Furthermore, the guidelines for systemic risk governance by the International Risk Governance Center (IRGC, Florin et al. 2018) also put strong emphasize on how individual failures may trigger systemic risks and gave recommendations on how to avoid them.

By recognizing that the failure of individual risk(s) may cause the realization of systemic risk, it is worthwhile to think about ways to combine them as both are intrinsically connected. However, while the relationship between individual failure and systemic risks is generally acknowledged how actually individual and systemic risks can be studied together is still a large question at hand; one possible way forward as proposed in this book is to treat a copula as a network property. The starting point of our discussion on how to integrate individual and systemic risk is Sklar's theorem which states that a multivariate distribution can be separated into two parts, one part describing the marginal distributions and one part describing the dependency between the distributions through a copula (Sklar 1959). Due to the central importance of Sklar's theorem, we present the theorem for the 2-dimensional case next.

Theorem 1.1 *Let $F_X \in R^2$ be a 2-dimensional distribution function with continuous marginals $F_{X_1}(x_1)$ and $F_{X_2}(x_2)$. Then there exists a unique copula C such that for all* x $\in R^2$

$$F_X(x_1, x_2) = C(F_{X_1}(x_1), F_{X_2}(x_2)) \tag{1.1}$$

Conversely, if C is a copula and $F_{X_1}(x_1)$ and $F_{X_2}(x_2)$ are distribution functions then the function F_X is a bivariate distribution with margins $F_{X_1}(x_1)$ and $F_{X_2}(x_2)$.

From a statistical standpoint, a copula can be interpreted as a multivariate distribution with univariate marginals. However, reinterpreting this theorem from a system's or network perspective, as it is done in this book (see Fig. 1.4), it is suggested that the (marginal) distributions (e.g., $F_{X_1}(x_1)$ and $F_{X_2}(x_2)$) can be treated as individual risks while the copula C models the interconnectedness between the individual risks (Hochrainer-Stigler et al. 2018b) in the system. Read in this way Sklar's theorem enables an integration of the measurement, modeling, and management of individual, including extreme risks, as well as systemic risks. For illustration purposes of this idea, we first want to introduce and discuss a two-node network example which is conceptually shown in Fig. 1.6.

Let's assume that each of the nodes in Fig. 1.6 is at risk and the individual risk can be represented through a loss distribution. From a network perspective, the system at hand is a two-node network and the copula approach establishes the non-linear dependence between the two nodes. As indicated in the Figure (see also Hochrainer-Stigler et al. 2018b), different strengths of connections between the two nodes can be modeled through the copula approach, for example, a loose connection between node 1 and node 2 during normal events and a tight connection between node 1 and node 2 during extremes. In more detail, during "normal" events (lower left corner of the scatterplot and marked in blue), and corresponding to the low severity events for node 1 and 2, the copula indicates a loose connection between them. With increasing severity and corresponding to the tails of the two nodes (low probability but high-consequence events, i.e., extremes), the copula model enforces a tighter connection between them (upper right corner of the scatterplot, marked in red). Dependent on the research domain this can be reinterpreted as a network property such as higher risk of spreading (e.g., diseases) or higher dependency of extreme outcomes (e.g., natural hazards, default of banks, total losses). Figure 1.6 also indicates that the copula has uniform marginals between [0, 1], so even small disturbances (e.g., compared to the aggregate absolute level) in one node can result in a tight connection with another node which subsequently can result in large overall disturbances.

We argue that from a network perspective Sklar's theorem enables the combination of the marginal distributions (i.e., individual risk) and the dependency structure of a system in a very convenient way as both can be analyzed independently as well as jointly. Consequently, this perspective enables an integration of individual and

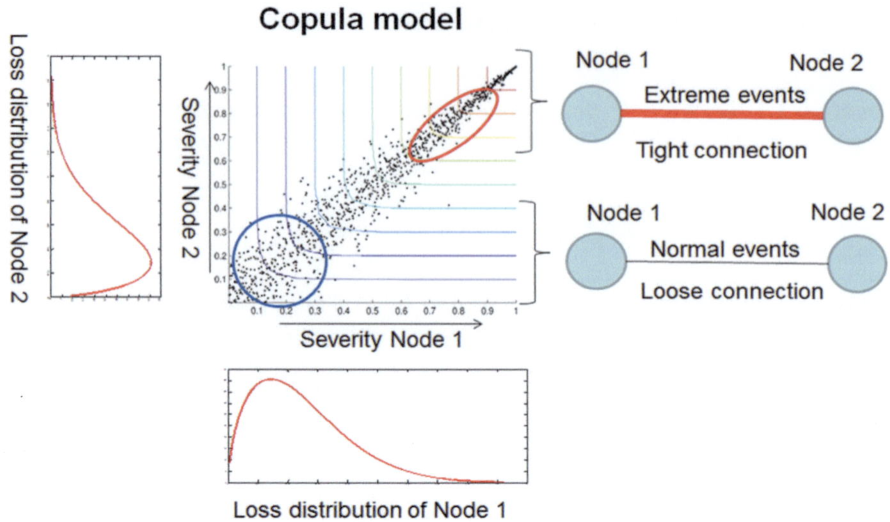

Fig. 1.6 Illustration of the copula concept as a network property. *Source* Based on Hochrainer-Stigler (2016) and Hochrainer-Stigler et al. (2018b)

systemic risk analysis within a common framework. For further illustration, we are providing a numerical example of the suggested approach from a decision-maker perspective.

Imagine that a decision-maker is responsible for a system which consists of two elements. It could also be imagined that one of the decision-makers of the two elements is willing to take a system perspective, e.g., as he is worried that in case of systemic risk realization he would be heavily affected. We take this consideration into account in Chap. 4 but here, for simplicity, let's assume that there is a decision-maker that is responsible for the whole system and two additional decision-makers which are responsible for each of the elements. These elements can be "at risk" due to exogenous shocks, e.g., due to a natural hazard event, which can cause losses. The decision-maker asks the two elements (we call the decision-makers of the elements for simplicity from now on nodes) in his system (from now on called network) to assess their risk. Let's assume that node 1 applies EVT and finds out that the risk can be represented through a Gamma distribution with shape parameter 1 and scale parameter 2. Therefore, the average annual loss for node 1 is 2, and the variance is 4. As node 1 is interested in tail events, it also looks at a possible 100-year event loss which it calculates to be 9.2. The decision-maker of node 1 gives this information to the decision-maker on the system level. The same procedure is performed by node 2 but here a Gamma distribution with shape parameter 7.5 and scale parameter 1 was estimated. The mean and variance are in this case, 7.5 and 7.5, respectively. The 100-year event loss would be 15.3. Hence, extreme losses are much higher for node 2 compared to node 1 and also the shape of the distribution is quite distinct (see the histograms on the left-hand side as well as below in Fig. 1.7). The data used by the two nodes is sent to the decision-maker and he looks a the individual distribution of losses as well relates the losses for the two nodes using a scatterplot as depicted in Fig. 1.7.

The decision-maker acknowledges that there is obviously some kind of relationship between the two nodes in respect to losses (see the scatterplot in Fig. 1.7). However, as there are different scales for the two nodes (see the x- and y-axes of the scatterplot) as well as quite different shapes for the two distributions, it is not immediately clear for him how an actual relationship between the two nodes and corresponding losses can be established. The decision-maker therefore wants to investigate this issue in more detail and tries to calculate the dependence of the two individual risks using a copula approach. He applies the corresponding techniques laid out in this book and finds out that a flipped Clayton copula with parameter 5 would be an appropriate model to take the dependencies explicitly into account. Figure 1.8 is showing the corresponding copula with univariate margins. Interestingly for him, he finds that there is a strong dependence in the tails but actually no strong dependencies in the region of frequent events. This is a quite different relationship than Fig. 1.7 would have initially suggested (see also the excellent example in Hofert et al. 2018).

Being satisfied with the copula approach to measure dependency and having the individual risks estimated he applies simulation techniques (which will be discussed in Chap. 3) to calculate a loss distribution on the system level. He finds out that a 100-year event on the system level would cause losses of about 22 currency units. Let's assume for the moment that he has a budget of 20 currency units available and in case of losses larger than this amount systemic risk would realize. Let's further assume that due to his risk aversion he wants to be saved for all frequent events up to the 100-year event. Obviously, we are oversimplifying what risk aversion is and how systemic risk events can be defined. We discuss these issues in full detail in the later chapters but here the main goal is to convey the message of the advantages of

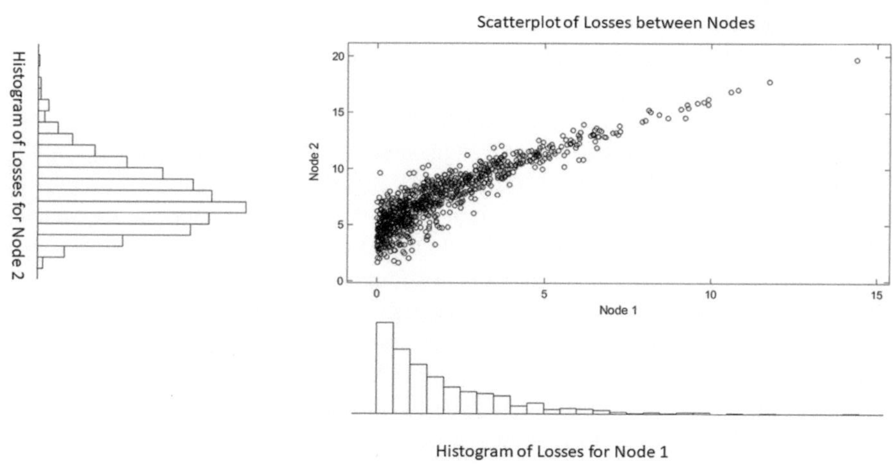

Fig. 1.7 Scatterplot of losses for the two nodes network example

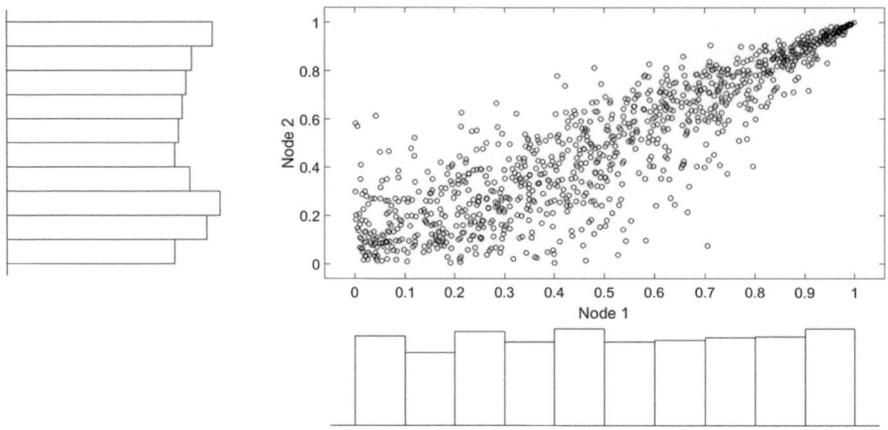

Fig. 1.8 Flipped Clayton copula and corresponding univariate marginals for the two nodes network example

integration. Given these assumptions, he wants to consider lowering the losses up to this risk level through decreasing the dependency between the nodes. In doing so, he uses the independence copula and performs the simulation again. For the independent case, i.e., no dependency between the nodes, he would be able to decrease the system level losses of a 100-year event down to 19 currency units which he would be therefore satisfied with. The decrease in the tail risk is quite significant (see Sect. 2.2). How he achieved independence and how much it would cost is irrelevant for our example here. Relevant is that he is able to decrease systemic risk through changes in the dependency structure. Note, he still has 1 currency unit available and can therefore use this amount to invest in the specific nodes, e.g., in the middle risk layer for node 2. This will further decrease systemic risk levels and additional investments can be done in an iterative manner till the available budget is depleted or an appropriate risk level on both (i.e., individual and system) levels for all decision-makers is reached. This ends our example.

Summarizing, we defined a system from a network perspective as a set of individual risks which are interconnected. We suggested to treat individual risks as random variables that can be represented through loss distributions. We further suggested that the interconnection between the individual risks in the system can be established through copulas. Through the copula approach, it is possible to derive a loss distribution on the system level as well. Systemic risk can therefore defined in a similar way as individual risk, e.g., through risk measures, however, it is important to note that the management of it would emphasize on the dependence while for individual risk one would emphasize on the risk layers (for a detailed and precise discussion see Chap. 3). As a consequence, it is possible to not only look at one particular scale but across multiple scales to determine individual and systemic risks. Indeed, what actually constitutes individual or systemic risk is dependent on the scale which one is looking from, e.g., systemic risk at one level may be just individual risk on another level (see Fig. 1.3). Consequently, our approach should enable an integrated and comprehensive analysis for both types of risk across a broad range of applications.

1.4 The Way Forward: Structure of Book and Related Literature

The organization of this book follows a structure similar to the introductory sections. In the next Chap. 2 we will discuss how to measure, model and manage individual and extreme risks using loss distributions as the central concept. Afterward, Chap. 3 presents how to measure, model and manage systemic risk through the use of copulas and how the integration of both can be established. Chapter 4 then shows various applications of our suggested approach on different scales focusing on natural disaster events such as floods and droughts. All of the chapters are written in such a way that they should be useful for a large audience. Most, if not all, topics only require a moderate level of knowledge of statistical and mathematical concepts on the undergraduate level. Furthermore, all techniques introduced and presented will be guided with examples and applications to increase understanding and proofs will be given only in some few cases. Furthermore, an extensive literature and discussion of it is given in each chapter for the interested reader. The goal of each chapter is to ensure that the reader is able to apply the presented techniques in real-world applications while he is able to find further reading for more advanced analysis. Many of the discussed concepts and discussions given in this book can also be found in other textbooks, for example, the seminal book of Embrechts et al. (2013) about extreme value theory (focusing on insurance and finance applications) and Nelsen (2007) or Salvadori et al. (2007) about copula approaches. The reinterpretation of Sklar's theorem as presented here was also already introduced in Hochrainer-Stigler (2016), Hochrainer-Stigler et al. (2018b). Additionally, the application examples given in Chap. 4 are based on research done by Lugeri et al. (2010), Jongman et al. (2014), Timonina et al. (2015b), Gaupp et al. (2017). Nevertheless, a comprehensive discussion of the suggested approach has not been presented yet, especially the techniques for both types of risk, and this is one main goal of the book: To promote the use of extreme and systemic risk analysis in an integrated way, here, using Sklar's theorem as one possible and promising avenue.

In doing so Chap. 2 will first focus on individual risk represented in the form of a cumulative distribution function and special emphasis will be given to pure downside risk, e.g., losses, represented through a so-called loss distribution. Section 2.1 gives an introduction into distributions, different families of distributions as well as how to estimate and model them. Particular emphasis is given on measures of risk including location, dispersion, and tail measures. As for extremes a theory of its own is needed the next section (Sect. 2.2) introduces extreme value theory and tools and methods to estimate accurately the tail of a loss distribution. This includes the classic block maxima and threshold models as well as point processes and special cases of it including k largest order statistics. In Sect. 2.3, we will discuss how such distributions can be used to determine risk management options, focusing on the certainty equivalent and the asking and bid price as well as the insurance premium for a risky situation. The Arrow-Lind theorem and its violation in case of extremes is looked at as well and a discussion of some risk management instruments currently employed in real-world applications is given at the end.

Moving from individual risk in Chap. 2 to dependency in Chap. 3 we first introduce the notion of a system within a network approach, i.e., a system will be defined as a set of interconnected elements that are "at risk". The interconnectedness will be viewed as dependency and various association measures are introduced in Sect. 3.1 especially the concept of a copula. As in the previous Chap. 2 also here measuring and modeling aspects will be introduced and examples are given. This will lead to Sect. 3.2 which is concerned about the structure of dependencies, and the focus will be given on so-called vine approaches as a way forward on how to determine the dependency structure through spanning trees. Having established all the necessary ingredients, Sect. 3.3 introduces Sklar's theorem and discusses its implication for integrating individual and systemic risks within a joint framework by reinterpreting Sklar's theorem from a network perspective. Again, measuring, modelling and management aspects will be presented and discussed based on some multi-node network examples. To the best knowledge of the author this perspective was first introduced in Hochrainer-Stigler (2016) and in more detail in Hochrainer-Stigler et al. (2018b). However, it should also be noted that the concept of a copula and its various applications are not new and all the techniques used and presented in the book are already well established in the respective fields, e.g., estimation of the marginals through extreme value theory and the copula structuring using vines. Nevertheless, the intent in this book is to present these concepts jointly together treating the copula as a network property. We conclude this chapter with a discussion on how to increase resilience for both types of risk and across scales in Sect. 3.4.

Afterwards, Chap. 4 presents specific case studies that use the approaches discussed before and should show the possible applications of such an integrative perspective focusing on different geographical levels and especially in the context of large-scale climate-related disaster events such as floods and droughts. The focal point in Sect. 4.1 is how to derive at loss distributions on larger scales taking dependencies explicitly into account. In the first application in Sect. 4.1.1 a rather simplified approach is presented, however, in cases of severe data scarcity it may be a first step worthwhile to be taken. Sect. 4.1.2 presents a more advanced approach and is applied to the European level with a special focus on climate change effects. The third application in Sect. 4.1.3 takes a look at drought events using the full set of sophisticated models to derive at loss distributions to be used also for systemic risk analysis. In Sect. 4.2, we will take a look at related measurement and risk management applications. Section 4.2.1 applies the risk layer approach to risk reduction and insurance-related instruments. Section 4.2.2 introduces a large-scale financing vehicle on the EU level, namely, the European Solidarity Fund. Additionally, multi-hazard risks on the global scale and related stress testing will be discussed in Sect. 4.2.3. As there are also some serious limitations of the suggested approach, we discuss in Sect. 4.3 ways forward within the context of agent-based modeling approaches, especially useful for the study of emergent behavior and cascading risks.

Finally, Chap. 5 discusses less quantitative aspects equally important for the management of individual and systemic risks, namely, human agency as well as governance aspects. Special focus will be given on recent analogies made between human systems and biological or ecological ones (Haldane and May 2011). As will be discussed, such analogies are dangerous as they can obscure the problems rather than illuminating them. The chapter ends with some suggestions of possible entry points on how human agency can be incorporated in the risk analysis of individual and systemic risk as well as related governance challenges drawing on lessons learned from related climate change governance processes in the past and recent climate risk management approaches recently suggested.

References

Abadie LM, Galarraga I, de Murieta ES (2017) Understanding risks in the light of uncertainty: low-probability, high-impact coastal events in cities. Environ Res Lett 12(1):014017

Barabási A-L, Albert R (1999) Emergence of scaling in random networks. Science 286(5439):509–512

Battiston S, Puliga M, Kaushik R, Tasca P, Caldarelli G (2012) Debtrank: too central to fail? Financial networks, the fed and systemic risk. Sci Rep 2:541

Benson C, Arnold M, de la Fuente A, Mearns R (2012) Financial innovations for social and climate resilience: establishing an evidence base. Social Resilience & Climate Change Brief, The World Bank, Washington, DC, USA

Centeno MA, Nag M, Patterson TS, Shaver A, Windawi AJ (2015) The emergence of global systemic risk. Annu Rev Sociol 41:65–85

Coles S, Bawa J, Trenner L, Dorazio P (2001) An introduction to statistical modeling of extreme values, vol 208. Springer

Collins A, Tatano H, James W, Wannous C, Takara K, Murray V, Scawthorn C, Mori J, Aziz S, Mosalam KM et al (2017) The 3rd global summit of research institutes for disaster risk reduction: expanding the platform for bridging science and policy making. Int J Disaster Risk Sci 8(2):224–230

Elias N (1987) Involvement and detachment. Basil Blackwell, Oxford

EM-DAT (2019) EM-DAT: emergency events database

Embrechts P, Klüppelberg C, Mikosch T (2013) Modelling extremal events: for insurance and finance, vol 33. Springer Science & Business Media

Florin M-V, Nursimulu A, Trump B, Bejtullahu K, Pfeiffer S, Bresch DN, Asquith M, Linkov I, Merad M, Marshall J et al (2018) Guidelines for the governance of systemic risks. Technical report, ETH Zurich

Folke C, Carpenter S, Walker B, Scheffer M, Elmqvist T, Gunderson L, Holling CS (2004) Regime shifts, resilience, and biodiversity in ecosystem management. Annu Rev Ecol Evol Syst 35:557–581

Frank AB, Collins MG, Levin SA, Lo AW, Ramo J, Dieckmann U, Kremenyuk V, Kryazhimskiy A, Linnerooth-Bayer J, Ramalingam B et al (2014) Dealing with femtorisks in international relations. Proc Natl Acad Sci 111(49):17356–17362

Gaupp F, Pflug G, Hochrainer-Stigler S, Hall J, Dadson S (2017) Dependency of crop production between global breadbaskets: a copula approach for the assessment of global and regional risk pools. Risk Anal 37(11):2212–2228
Haldane AG, May RM (2011) Systemic risk in banking ecosystems. Nature 469(7330):351
Haraguchi M, Lall U (2015) Flood risks and impacts: a case study of thailand's floods in 2011 and research questions for supply chain decision making. Int J Disaster Risk Reduct 14:256–272
Helbing D (2013) Globally networked risks and how to respond. Nature 497(7447):51
Helbing D (2018) Towards digital enlightenment: essays on the dark and light sides of the digital revolution. Springer
Heppenstall AJ, Crooks AT, See LM, Batty M (2011) Agent-based models of geographical systems. Springer Science & Business Media
Hochrainer S (2006) Macroeconomic risk management against natural disasters. German University Press
Hochrainer S, Mechler R, Pflug G, Lotsch A (2008) Investigating the impact of climate change on the robustness of index-based microinsurance in Malawi. The World Bank
Hochrainer-Stigler S (2016) Systemic and extreme risks: ways forward for a joint framework. In: IDRiM 2016 7th international conference on integrated disaster risk management disasters and development: towards a risk aware society, October 1–3, 2016. Islamic Republic of Iran, Isfahan
Hochrainer-Stigler S, Keating A, Handmer J, Ladds M (2018a) Government liabilities for disaster risk in industrialized countries: a case study of Australia. Environ Hazards 17(5):418–435
Hochrainer-Stigler S, Pflug G, Dieckmann U, Rovenskaya E, Thurner S, Poledna S, Boza G, Linnerooth-Bayer J, Brännström Å (2018b) Integrating systemic risk and risk analysis using copulas. Int J Disaster Risk Sci 9(4):561–567
Hochrainer-Stigler S, Colon C, Boza G, Brännström Å, Linnerooth-Bayer J, Pflug G, Poledna S, Rovenskaya E, Dieckmann U (2019) Measuring, modeling, and managing systemic risk: the missing aspect of human agency. J Risk Res, 1–17
Hofert M, Kojadinovic I, Mächler M, Yan J (2018) Elements of copula modeling with R. Springer
Holling CS (1973) Resilience and stability of ecological systems. Annu Rev Ecol Syst 4(1):1–23
Jongman B, Hochrainer-Stigler S, Feyen L, Aerts JC, Mechler R, Botzen WW, Bouwer LM, Pflug G, Rojas R, Ward PJ (2014) Increasing stress on disaster-risk finance due to large floods. Nat Clim Change 4(4):264
Kovacevic RM, Pflug G (2015) Measuring systemic risk: structural approaches. Quant Financ Risk Manag Theory Pract, 1–21
Linnerooth-Bayer J, Hochrainer-Stigler S (2015) Financial instruments for disaster risk management and climate change adaptation. Clim Change 133(1):85–100
Little RG (2002) Controlling cascading failure: understanding the vulnerabilities of interconnected infrastructures. J Urban Technol 9(1):109–123
Lugeri N, Kundzewicz ZW, Genovese E, Hochrainer S, Radziejewski M (2010) River flood risk and adaptation in Europe-assessment of the present status. Mitig Adapt Strat Glob Change 15(7):621–639
Luhmann N (1995) Social systems. Stanford University Press
Luhmann N, Baecker D, Gilgen P (2013) Introduction to systems theory. Polity Cambridge
McNeil AJ, Frey R, Embrechts P (2015) Quantitative risk management: concepts, techniques and tools-revised edition. Princeton University Press
Mechler R (2016) Reviewing estimates of the economic efficiency of disaster risk management: opportunities and limitations of using risk-based cost-benefit analysis. Nat Hazards 81(3):2121–2147
Mechler R, Bouwer LM, Linnerooth-Bayer J, Hochrainer-Stigler S, Aerts JC, Surminski S, Williges K (2014) Managing unnatural disaster risk from climate extremes. Nat Clim Change 4(4):235
Miller MB (2018) Quantitative financial risk management. Wiley
Munich Re (2018) NatCatSERVICE methodology. Munich Re, Munich
Nelsen RB (2007) An introduction to copulas. Springer Science & Business Media

Nicholson S, Klotter D, Chavula G (2014) A detailed rainfall climatology for Malawi, Southern Africa. Int J Climatol 34(2):315–325

Pflug G, Roemisch W (2007) Modeling, measuring and managing risk. World Scientific

Poledna S, Thurner S (2016) Elimination of systemic risk in financial networks by means of a systemic risk transaction tax. Quant Financ 16(10):1599–1613

Sachs L (2012) Applied statistics: a handbook of techniques. Springer Science & Business Media

Salvadori G, De Michele C, Kottegoda NT, Rosso R (2007) Extremes in nature: an approach using copulas, vol 56. Springer Science & Business Media

Scheffer M, Carpenter SR (2003) Catastrophic regime shifts in ecosystems: linking theory to observation. Trends Ecol Evol 18(12):648–656

Sklar M (1959) Fonctions de repartition an dimensions et leurs marges. Publ Inst Statist Univ Paris 8:229–231

SREX (2012) Special report of working groups i and ii of the intergovernmental panel on climate change (IPCC). Cambridge University Press

Thurner S, Hanel R, Klimek P (2018) Introduction to the theory of complex systems. Oxford University Press

Timonina A, Hochrainer-Stigler S, Pflug G, Jongman B, Rojas R (2015a) Deliverable 3.3: probabilistic risk assessment of high impact events to be used in the case studies of enhance. Deliverable 3.3, ENHANCE Project, Brussels

Timonina A, Hochrainer-Stigler S, Pflug G, Jongman B, Rojas R (2015b) Structured coupling of probability loss distributions: assessing joint flood risk in multiple river basins. Risk Anal 35(11):2102–2119

United Nations (2015a) The sustainable development goals. United Nations

United Nations (2015b) Sendai framework for disaster risk reduction 2015–2030. United Nations

United Nations (2016) Report of the open-ended intergovernmental expert working group on indicators and terminology relating to disaster risk reduction. United Nations

Wallemacq P, House R (2018) UNISDR and CRED report: economic losses, poverty & disasters (1998–2017)

Winsemius HC, Aerts JC, van Beek LP, Bierkens MF, Bouwman A, Jongman B, Kwadijk JC, Ligtvoet W, Lucas PL, Van Vuuren DP et al (2016) Global drivers of future river flood risk. Nat Clim Change 6(4):381

Chapter 2
Individual Risk and Extremes

In Chap. 1, we introduced the overall approach on how to integrate extreme and systemic risk analysis. We discussed that a network perspective is very beneficial in that regard. We informally defined a system to be a set of individual elements which are, at least partly, interconnected with each other. We assumed that these individual elements are "at risk" and we will call such kind of risk "individual risk". In this chapter, we assume that the risk an individual element within a network is exposed to can be represented as a random variable. Consequently, our focal point for individual risk will be a distribution function. As we are especially interested in downside risk (e.g., losses), we will frequently deal with a so-called loss distribution (Hogg and Klugman 2009) which is discussed in Sect. 2.1 first. For extreme risk, the tails of the distribution function are of great importance. However, for accurately estimating the tail, a theory of its own as well as corresponding techniques are needed (Malevergne and Sornette 2006). It is therefore necessary to discuss extreme value theory and extreme value statistics in some detail, which we do in Sect. 2.2. Finally, in Sect. 2.3 we relate individual risk, including extremes, with decision theory for determining appropriate risk management options to decrease risk to acceptable levels.

2.1 The Loss Distribution

Generally speaking, there are different ways how individual risk can be measured, modeled, and managed. Regarding risk analysis of individual risk from a quantitative perspective, probability-based approaches are superior compared to single or scenario-based analyses as all possible events that could happen can be looked at (Pflug and Roemisch, 2007). Consequently, not only different risk measures can be employed but also management approaches can be adapted to the specific interest of a decision-maker, e.g., focusing on frequent events by using location measures or extreme events using tail measures (Mechler et al. 2014). We first start with some definitions and more technical details of distributions to set up the stage on how such

© Springer Nature Singapore Pte Ltd. 2020

S. Hochrainer-Stigler, *Extreme and Systemic Risk Analysis*, Integrated Disaster Risk Management, https://doi.org/10.1007/978-981-15-2689-3_2

distributions are useful for measuring specific aspects of risk, including extremes, and how they can be beneficial for risk management purposes. We assume that the reader is familiar with some basic facts about random variables and distributions in general, and in case not we refer to the various statistical textbooks available, for example, Klugman et al. (2012) for an introduction to loss models.

The modeling of random quantities (such as the magnitude of natural hazard events or corresponding losses, which are intrinsically random) as random variables needs the concept of a probability space $(\Omega, \mathscr{F}, \mathscr{P})$. It includes a scenario set (sometimes also called the sample space) Ω of all possible outcomes, a σ-algebra on Ω, i.e., a family of subsets of Ω which are satisfying three axioms ($\Omega \in \mathscr{F}$, if $A \in \mathscr{F}$ its complement $\Omega \setminus A$ also belongs to $\mathscr{F}$, and every countable union of sets in $\mathscr{F}$ also belongs to $\mathscr{F}$), and a probability measure $\mathscr{P}$ which also satisfies certain properties ($\mathscr{P}(\Omega) = 1$, and $\mathscr{P}$ is countable additive). For the mathematical treatment of random variables within Measure Theory, we refer to the classic book of Bauer (2011) for a comprehensive introduction. In this book, we assume that the sample space is either discrete or continuous and the σ-Algebra is the power set of Ω or in the continuous case the Borel σ-algebra.

A random variable X is then a real-valued function defined on a probability space and characterized by its cumulative distribution function F with $F(x) = P(X \leq x)$. In other words, the cumulative distribution function (or simply distribution) of a random variable X is the probability that X is less than or equal to a given number (e.g., in Fig. 2.1, $P(X \leq x_1)$). The cumulative distribution function F has several properties worth mentioning (and is related to a probability measure). It is non-decreasing and right-continuous; the values of $F(x)$ are between 0 and 1 for all x $\lim_{x \to -\infty} F(x) = 0$ and $\lim_{x \to \infty} F(x) = 1$. Note, $F(x)$ need not be left-continuous therefore it is possible for the distribution to jump. As already indicated above, discrete and continuous types of distributions are usually distinguished, however, it should also be noted that Measure Theory and (Lebesgue) Integration provides a theory so that such a

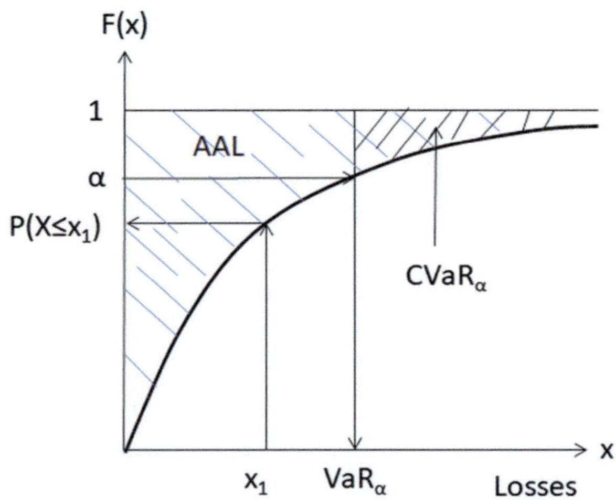

Fig. 2.1 Example of a distribution function and various risk measures

separation is actually not needed (Bauer 2011). It is not our intention to go into more details here as most of the models for random variables considered will have a (absolute) continuous distribution function and therefore possess densities, notated as $f(x)$. In more detail, a distribution has a density function $f(x)$ if and only if its cumulative distribution function $F(x)$ is absolutely continuous. In this case F is almost everywhere differentiable, and its derivative can be used as a probability density which is the derivative of the cumulative distribution function, i.e.,

$$\frac{d}{dx}F(x) = f(x).$$

Some of the most common distribution functions are discussed next. We are starting with the normal distribution as it is probably the most famous distribution of all distributions because it fits many natural phenomena. Suppose X is a continuous random variable having the density

$$f(x) = \frac{1}{\sqrt{2\pi\sigma^2}} \exp\left(-\frac{(x-\mu)^2}{2\sigma^2}\right) \qquad -\infty < x < \infty$$

then X is said to be a normal distributed random variable with parameters $\mu \in \mathbb{R}$ and $\sigma^2 > 0$, abbreviated as $X \sim N(\mu, \sigma^2)$. In case X is a continuous random variable with density

$$f(x) = \frac{x^{n/2-1}e^{-x/2}}{2^{n/2}\Gamma(n/2)} \qquad x \geq 0$$

then one says that X has a χ^2 distribution with $n \in \mathbb{N}$ degrees of freedom, abbreviated as $X \sim \chi^2(n)$. The Gamma function $\Gamma(\alpha)$ used in the formula above can be defined as

$$\Gamma(\alpha) = \int_0^\infty x^{\alpha-1}e^{-x}dx.$$

In the case that $Z \sim N(0, 1)$ and $U \sim \chi^2(n)$ and given they are independent from each other, one can define a random variable X as

$$X = \frac{Z}{\sqrt{\frac{U}{n}}},$$

which has the following density function

$$f(x) = \frac{\Gamma(\frac{n+1}{2})}{\sqrt{\pi n}\Gamma(\frac{n}{2})}\left(1 + \frac{x^2}{n}\right)^{-\frac{n+1}{2}} \qquad -\infty < x < \infty$$

and one calls it the t-distribution, abbreviated as $X \sim t(n)$. A random variable X is said to follow the Beta distribution with parameters $\alpha > 0$ and $\beta > 0$ if

$$f(x) = \frac{\Gamma(\alpha+\beta)}{\Gamma(\alpha)\Gamma(\beta)} x^{\alpha-1}(1-x)^{\beta-1} \qquad x \in [0,1].$$

Usually, one writes this as $X \sim Beta(\alpha, \beta)$. We also want to mention the log normal distribution with density

$$f(x) = \frac{1}{\sqrt{2\pi}\sigma x} e^{-(\ln(x)-\mu)^2/2\sigma^2} \qquad x \geq 0$$

and parameters $\mu \in \mathbb{R}$ and $\sigma > 0$, usually notated as $X \sim lognormal(\mu, \sigma)$. Finally, and very important throughout the book, is the Gamma distribution. X is said to have a Gamma distribution with parameters $\alpha > 0$ and $\beta > 0$ (the shape and the rate parameter), notated as $X \sim \Gamma(\alpha, \beta)$, if it has density

$$f(x) = \frac{\beta^\alpha}{\Gamma(\alpha)} x^{\alpha-1} e^{-\beta x}$$

for $x > 0$ and zero otherwise. It should be noted that many more distribution functions exist and we refer to Sachs (2012) as well as Roussas (1997) for a comprehensive list.

Given that a distribution is selected as a possible candidate to specify the random variable X there is the question how the parameters of the underlying distribution can be estimated. Many different approaches for parameter estimation were developed, most prominently the maximum-likelihood method, but there are also other techniques available such as the L-moment methods as well as the method of moments (see Klugman et al. 2012). The maximum-likelihood method will be described in more detail in the next Sect. 2.2 and is therefore skipped here. Given that parameters are estimated with one of the available approaches for some candidate distributions (usually it is not known beforehand which distribution actually should be chosen) some Goodness of Fit (GoF) tests need to be performed to select the most appropriate one. Especially three GoF statistics are used very often in practice including the Kolmogorov-Smirnov, the Anderson-Darling, and the χ^2 test. In addition, there are also some other more broad-based information criteria available for model selection such as the Akaike Information Criterion (AIC) and the Bayesian Information Criterion (BIC) but will be discussed later in Sect. 3.1 in more detail. Here, due to their importance the three GoF test statistics will be shortly presented next. The basic idea is quite similar for all three approaches, namely, to focus on the difference between the fitted and the observed empirical distribution.

The Kolmogorov-Smirnov (KS) test focuses on the difference between the empirical and fitted cumulative distribution. Hence, it can be used to test if a given sample stems from a population with a specific distribution. The empirical distribution function $F_n(x)$ for observations x_i, $i = 1, \ldots, n$ (n is the total number of observations) and all observations assumed to be independent and identically distributed (i.i.d.) is defined as

$$F_n(x) = \frac{1}{n}\sum_{i=1}^{n} I_{[\infty,x]}(x_i),$$

where $I_{[\infty,x]}$ is the indicator function, i.e., $I_{[\infty,x]}(x_i) = 1$ if $x_i \in [\infty, x]$ and else 0. The Kolmogorov-Smirnov statistic is given by

$$D_n = sup_x|F_n(x) - F(x)|.$$

Under the null hypothesis (i.e., the sample is drawn from the fitted candidate distribution), the test statistic D_n converges to 0 almost surely with increasing sample size. Note, the KS test is influenced by the center of the distribution the most and less by the tails (see also the discussion of the properties of extremes in the next section) which may not always be appropriate.

The Anderson-Darling (AD) test can be seen as a refinement of the KS test as also some weight is given to the deviation in both tails of the distribution. Given the total number of observations again to be notated as n and the fitted cumulative distribution function denoted as $F(x)$ the AD statistic is given by

$$A_n^2 = -n - \sum_{i=1}^{n} \frac{2i-1}{n}[ln(F(y_i)) + ln(1 - F(y_{n+1-i}))],$$

where y_i are the ordered data. The smaller the value of the AD test statistic, the better the fit of the hypothetical distribution to the sample data.

The third test considered here is the χ^2 test which weights the center and the tails of the distribution equally. The χ^2 test can be applied to binned data and the test statistic is given by

$$\chi^2 = \sum_{i=1}^{n} \frac{(O_i - E_i)}{E_i},$$

where O_i being the number of observations in bin i and E_i being the number of expected data points in bin i. There are also many other GoF measures available and we refer to Panjer (2006) for a comprehensive discussion for the interested reader.

Throughout this chapter, we will give examples of each technique introduced using rainfall data from a weather station in Malawi (in Chitedze) where rainfall was measured from 1961 till 2005 on a daily basis (see Hochrainer et al. 2008). Figure 2.2 shows the average rainfall amount over the dekads (10-day periods, where each month has three dekads, hence, 36 dekads in total) starting from the beginning of October (i.e., dekad 1) till the end of September (dekad 36) the next year. As can be seen the rainy season is from October (starting with dekad 1) till March (ending

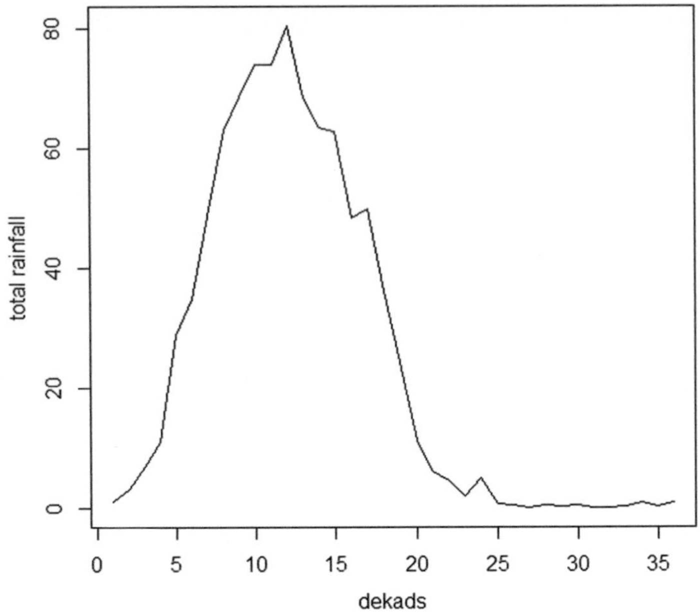

Fig. 2.2 Mean total rainfall in dekads from October (dekad 1) till September (dekad 36) in Chitedze, Malawi. *Source* Data used from Hochrainer et al. (2008)

with dekad 18) (see also Nicholson et al. 2014) and we focus on this period for our analysis.

As rainfall is often modeled through a Gamma distribution, we will apply a maximum-likelihood method to estimate corresponding parameters of rainfall for this period. It should be noted that it is also possible to further refine the rainy season into an early and late one (Nicholson et al. 2014) or looking at cumulative rainfall over specific dekads which would lead to different estimates, however, our focus will be on the full rainy season which is beneficial for the demonstration of the block maxima and threshold approach discussed in Sect. 2.2. For example, in Chap. 1 we already estimated the Gamma distribution for the dekad rainfall over the rainy season using the Method of Moments. It is also possible to apply the maximum-likelihood method as well, however, one has to make sure that there are no zeros in the data. The Kolmogorov-Smirnov and Anderson-Darling test as mentioned above indicate a reasonable fit. Hence, we would chose the estimated Gamma distribution for modeling dekad rainfall in the rainy season. However, as graphical inspections indicate (Fig. 2.3), especially the tails of the distribution are not fitted very well. We will come back to this point a little bit later in our discussion.

Given an appropriate distribution was found and parameters estimated the next question is what to do with it. A distribution is actually a complex object and there is the need to describe it with few relevant parameters. The actual decision which parameters should be chosen very much depends on what type of risks a decision-maker

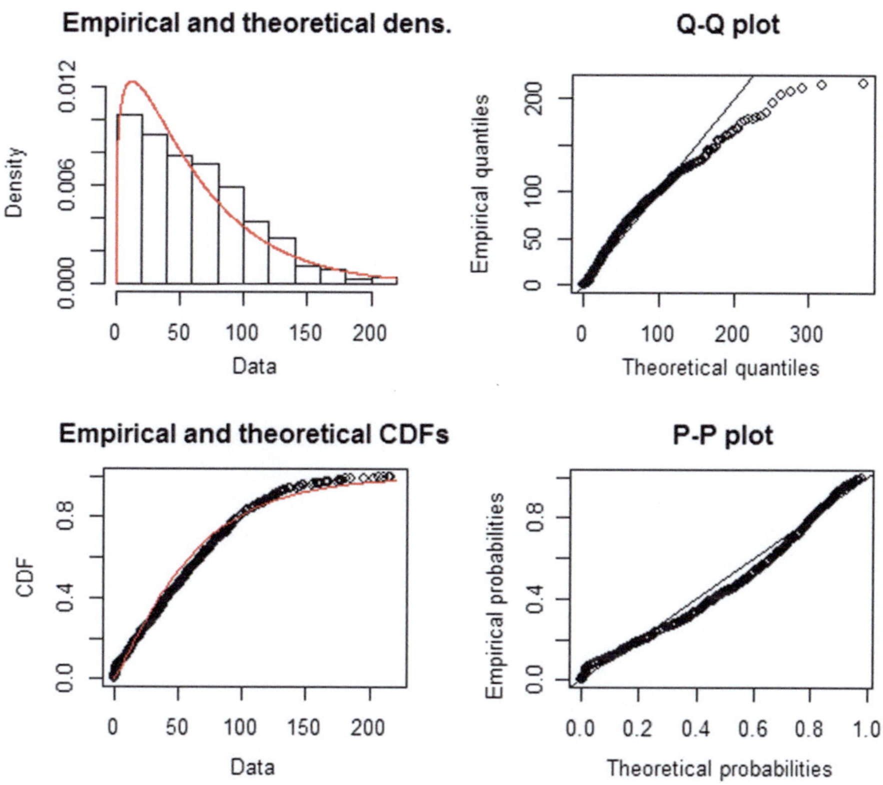

Fig. 2.3 Diagnostic check of the Gamma distribution model for dekad rainfall

is interested in (Linnerooth-Bayer and Hochrainer-Stigler, 2015). Most important in our context are location, dispersion, and tail measures. Dependence measures will be introduced in the next chapter within the discussion of copulas. Frequently used location measures include the *expectation*, the *median* and *mode*. The most prominent dispersion measures include the *variance*, the *standard deviation* as well as the *mean absolute deviation*. The expectation, median, and mode are used in many applications that are especially interested in the average behavior of a random variable and the variance and standard deviation are the primary risk measures for the behavior (e.g., fluctuation) around the mean (Mechler 2016). All of them more or less focus on the area of the distribution with the highest density (see Fig. 1.1). However, in recent years there has been a growing interest in risk measures that focus on the tails of a distribution, such as the *Value at Risk* or the *Expected Shortfall* sometimes also called *Conditional Value at Risk* (CVaR, see Fig. 2.1). A summary of some important risk measures is given in the Table 2.1.

For many distributions the corresponding risk measures listed in Table 2.1 can be expressed analytically. If not possible one still can apply sampling approaches to get approximate results. Very often it is easier to use the density function of the

Table 2.1 Selected location, dispersion and tail measures for unimodal absolute continuous distributions

Risk measure	Formula
Expectation $\mathbf{E}(X)$	$\int_{-\infty}^{+\infty} x dF(x)$
Median $\mathbf{Med}(X)$	$[F^{-1}(0.5) + F^{-1}(0.5+)]/2$
Variance σ^2	$\mathbf{E}(X - \mathbf{E}(X)^2)$
Standard Deviation $\mathbf{Std}(X)$	$\sqrt{\sigma^2}$
Value at Risk $\mathbf{V}aR_\alpha$ of level α	$F^{-1}(\alpha)$
Expected Shortfall ES	$\frac{1}{\alpha}\int_0^\alpha F^{-1}(x)dx$
Semi-Variance σ_t^2	$\int_{-\infty}^{t}(x-t)^2 f(x)dx$

distribution to derive at the risk measures or to perform simulations. As one example we take a look at the Gamma distribution which was already introduced and which is frequently used in many applied research areas due to its flexibility (e.g., distribution shapes, see, for example, the discussion in Wilks 1990). For illustration purposes on how to derive explicitly at some location and dispersion measures, we want to compute the kth moment of the Gamma distribution.

$$
\begin{aligned}
\mathbb{E}X^k &= \int_0^\infty x^k \frac{\beta^\alpha}{\Gamma(\alpha)} x^{\alpha-1} e^{-\beta x} dx \\
&= \frac{\beta^\alpha}{\Gamma(\alpha)} \int_0^\infty x^{(\alpha+k)-1} e^{-\beta x} dx \\
&= \frac{\beta^\alpha}{\Gamma(\alpha)} \frac{\Gamma(\alpha+k)}{\beta^{\alpha+k}} \int_0^\infty \frac{\beta^{\alpha+k}}{\Gamma(\alpha+k)} x^{(\alpha+k)-1} e^{-\beta x} dx \\
&= \frac{\beta^\alpha}{\Gamma(\alpha)} \frac{\Gamma(\alpha+k)}{\beta^{\alpha+k}} = \frac{\Gamma(\alpha+k)}{\Gamma(\alpha)\beta^k} \\
&= \frac{(\alpha+k-1)\Gamma(\alpha+k-1)}{\Gamma(\alpha)\beta^k} \\
&= \frac{(\alpha+k-1)(\alpha+k-2)\ldots\alpha\Gamma(\alpha)}{\Gamma(\alpha)\beta^k} \\
&= \frac{(\alpha+k-1)\ldots\alpha}{\beta^k}.
\end{aligned}
$$

Note, the last integral integrates to 1 as it is the density function of a Gamma distribution with parameter $\alpha + k$ and β. Having the kth moment it is easy to derive at the expectation and variance. For example, the expectation is $\mathbf{E}X = \frac{\alpha}{\beta}$ and the variance can be derived through $\boldsymbol{\sigma}^2(X) = \mathbf{E}X^2 - (\mathbf{E}X)^2 = \frac{(\alpha+1)\alpha}{\beta^2} - (\frac{\alpha}{\beta})^2 = \frac{\alpha}{\beta^2}$. Note, the expectation can be also calculated as the area above the curve of a distribution such as shown in Fig. 2.1.

For tail measures such as the Value at Risk one can simply include the α value in the inverse of the distribution (see again Fig. 2.1). The generalized inverse of the distribution function F will be introduced in the next chapter in more detail. Simply put $x_\alpha = F^{-1}(\alpha)$ defines the α-quantile of F and for illustration we introduce the standard exponential function $F = 1 - exp(-x)$ for $x \geq 0$ and zero otherwise. This is also a continuous distribution and also used throughout the book. The inverse of the standard exponential distribution is

$$\begin{aligned} F(x) &= 1 - exp(-x) \\ exp(-x) &= 1 - F(x) \\ x &= -log(1 - F(x)) \\ x_\alpha &= F^{-1}(\alpha) = -log(1 - \alpha). \end{aligned}$$

So, in this case, the Value at Risk can be quite easily calculated for any α level. In case no analytical results for the location and dispersion measures are available one can also use the inverse of the distribution and apply the inverse transformation method. This is a simple method for simulating a random variable X distributed as F.

Theorem 2.1 *Let $F(x)$ be a distribution function. Let $F^{-1}(y), y \in [0, 1]$ denote the inverse function. Define $X = F^{-1}(U)$, where U is uniformly distributed over [0,1]. Then X is distributed as F.*

Therefore, one can also simulate a large sample using the above theorem to estimate the expectation and variance through the use of sample statistics. The theorem is also important for modeling purposes as will be discussed in later chapters. Within catastrophe risk modeling instead of the loss distribution, sometimes also the so-called exceedance probability curve is used, i.e., instead of $P(X \leq x)$ one uses $P(X \geq x)$. The measures presented above would need to be adapted to this case but there is actually no reason to use an exceedance probability curve instead of the distribution function. Because the later is the original and the former mainly used within catastrophe modeling we focus on the distribution throughout the book.

Having defined and established the notion of a (loss) distribution and measures on how to characterize specific aspects of the distribution, the next question is how to estimate such distributions using past observations. The starting point is usually that one tries to fit an appropriate distribution to the entire dataset of past observations using maximum-likelihood techniques and goodness of fit statistics to determine the best model. The method for performing such an analysis is implemented in many different software systems and the theory and application is standard and can be found in many statistical textbooks (see, for example, Sachs 2012). Such an approach is most suitable for non-extreme events. However, with such approaches one may underestimate extreme events that lie on the tail of the distribution and there is the question how such extreme events need to be dealt with. As we will see the answer is through extreme value theory.

Surprisingly, the distribution of certain rare events is largely independent of the distribution from which the data is drawn. Specifically, the maximum of a series of i.i.d. observations can take only one of three limiting forms (if at all). Hence, a single distribution model chosen for its overall fit to all historical observations, e.g., losses, may not provide a particularly good fit to extremes, e.g., large losses, and through extreme value theory one can accurately estimate the tail of a distribution. For example, a Gamma distribution may be the best model for the non-extreme cases (using the whole dataset for estimation) and a Gumbel distribution may be more appropriate for the tails of the distribution (using only past extreme observations for the estimation). The corresponding techniques which need to be applied will be discussed next in some detail.

2.2 Extreme Value Theory and Statistics

In this section most prominent statistical techniques to estimate the tails of distributions based on past observations are given. The underlying theory needed for this task is called Extreme Value Theory (EVT). The structure of this section is based on Coles et al. (2001) and while the underlying theory is presented in the necessary detail we omit the more theoretical aspects and refer the interested reader to Embrechts et al. (2013) and Reiss et al. (2007). EVT deals with the stochastic behavior of the maximum (or minimum) of i.i.d random variables. Contrary to location (e.g., mean, median) and deviation measures (e.g., variance) which focus on the center of the distribution, the distributional properties of extremes are determined by the upper and lower tails of the underlying distribution. As the name extremes already indicates, one major challenge in estimating accurately the tail of the distribution is data scarcity (see Fig. 1.1). Most data is (naturally) concentrated toward the center of the distribution and so, by definition, extreme data is scarce and therefore estimation is difficult. Furthermore, while there are very few observations in the tail, there are often estimates required beyond the largest observed data value. Importantly, while standard density estimation techniques fit well where data has greatest density, it can be severely biased in estimating tail probabilities. Hence, a theory of its own is needed for extremes (Malevergne and Sornette 2006).

Historically, work on extreme value problems may be traced back to 1709 when Bernoulli discussed the mean largest distance from the origin given n points lying at random on a straight line of a fixed length t. However, the foundations of the asymptotic argument which can be seen as an analogy to the central limit theorem were set out by Fisher and Tippett in the 1920's (Fisher and Tippett 1928). A rigorous foundation of extreme value theory was presented in 1943 by Gnedenko (1943). Statistical applications were first studied and formalized by Gumbel in the 1950's (and Jenkinson) (Gumbel 1958; Jenkinson 1955). The classical limit laws were generalized in the 1970s by Pickands (Pickands III et al. 1975).

One natural way to characterize the behavior of extremes is by considering the behavior of the sample maxima

$$M_n = \max\{X_1, X_2, ..., X_n\},$$

where $X_1, ..., X_n$ is a sequence of independent random variables having a common distribution function F. Then the cumulative distribution function of M_n is

$$P(M_n \leq x) = P(X_1 \leq x, ..., X_n \leq x) = [F(x)]^n.$$

In practice the distribution F is unknown and the approach to obtain an extreme value distribution is based on an asymptotic argument. Observe, as n tends to infinity for any fixed value of x the distribution of M_n converges to a degenerate distribution (e.g., as $F(x)$ is always smaller or equal to 1). The same problem of degeneracy also arises in other important limit theorems, e.g., the central limit theorem, and a linear transformation is needed to overcome this problem.

The distribution of maximas and the distribution of exceedances will provide the basis for modeling and estimating the tail behavior of real-world data. Basically, there are two main kinds of modeling approaches for extreme values: the (traditional) *block maxima* models and (more modern) models for *threshold exceedances*. The different approaches in selecting maxima are visualized in Fig. 2.4. While in the block maxima approach only the maximum in each block is taken for estimating tail parameters, e.g., X_2, X_5, X_7, X_{12}, threshold approaches look at extremes which are defined to be valued above some given threshold level, say u, e.g., $X_2, X_3, X_7, X_8, X_9, X_{12}$. Both approaches have respective advantages and disadvantages. We first introduce the

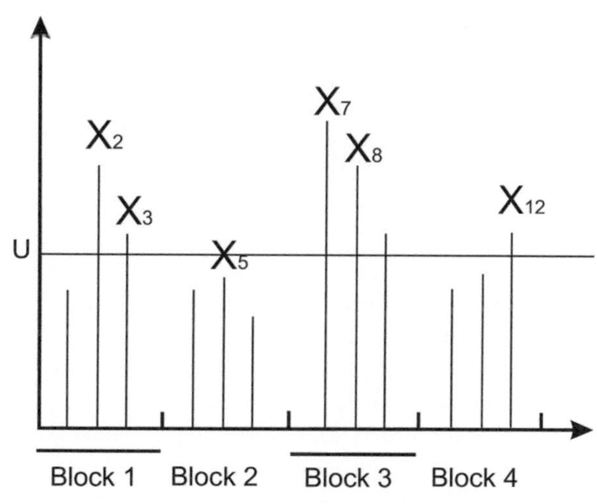

Fig. 2.4 Illustration of the block maxima and the threshold approach

block maxima approach and present the threshold approach afterwards. Following this discussion we also provide an unifying approach based on point processes, discuss related order statistics, and finally present issues in regard to short- and long-run dependence as well as non-stationarity.

2.2.1 Distributions of Maxima

As indicated before, to obtain a nondegenerate limiting distribution of the sample maxima it is necessary to "reduce" the actual greatest value by applying a linear transformation with coefficients that depend on the sample size n but not on x. For moving forward we need the concept of convergence in distribution which is usually notated as $\xrightarrow{d}$. Formally, a sequence $X_1, X_2, \ldots$ of random variables converge in distribution to a random variable X if $\lim_{n\to\infty} F_n(x) = F(x)$, for every number x$\in \mathbb{R}$ where F is continuous. Fisher and Tippett (1928) have shown that if a nondegenerate distribution for the maxima exists, it can only be one of three types. Hence, the following result is *the* basis of classic extreme value theory.

Theorem 2.2 *Let* (X_n) *be a sequence of i.i.d. r.v. If there exist norming constants* $c_n > 0, d_n \in \mathbb{R}$ *and some nondegenerate distribution function* H *such that*

$$\frac{M_n - d_n}{c_n} \xrightarrow{d} H,$$

then H belongs to the type of one of the following three distribution functions:

$$\text{Type 1 (Gumbel-type): } \Lambda(x) = \exp\{-e^{-x}\}, \qquad x \in \mathbb{R} \tag{2.1}$$

$$\text{Type 2 (Fréchet-type): } \Phi_\alpha(x) = \begin{cases} 0 & \text{if } x \le 0 \\ \exp\{-x^{-\alpha}\} & \text{if } x > 0 \end{cases} \quad \alpha > 0 \tag{2.2}$$

$$\text{Type 3 (Weibull-type): } \Psi_\alpha(x) = \begin{cases} \exp\{-(-x)^{\alpha}\} & \text{if } x \le 0 \\ 1 & \text{if } x > 0 \end{cases} \quad \alpha > 0 \tag{2.3}$$

.

Theorem 2.2 is often called the Fisher-Tippett theorem. Observe, the theorem does not guarantee the existence of a nondegenerate limit for M_n, nor does it specify which limit will arise when such a limit does exist. However, when such a limit does exist, the limiting distribution of sample maxima follows one of the distributions specified in the theorem.

As an example lets look again at the standard exponential distribution already introduced in Sect. 2.1 and assume the norming constants (see, for example, Kotz and Nadarajah 2000) to be $c_n = ln(\text{n})$ and $d_n = 1$. Then

$$
\begin{aligned}
P\left(\frac{M_n - c_n}{d_n} \leq x\right) &= P(M_n \leq d_n x + c_n) \\
&= P(X_1 \leq d_n x + c_n, \ldots, X_n \leq d_n x + c_n) \\
&= (P(X \leq d_n x + c_n))^n \\
&= (P(X \leq x + ln(n)))^n \\
&= (1 - exp(-x - ln(n)))^n \\
&= \left[1 - \frac{exp(-x)}{n}\right]^n \\
&\rightarrow exp(-exp(-x)) \qquad \text{as} \qquad n \rightarrow \infty.
\end{aligned}
$$

Hence, the maxima of a standard exponential distribution, if appropriately normalized, will follow (in the limit) the Gumbel distribution. For other examples, we refer to Kotz and Nadarajah (2000).

As the distributions are continuous the densities of each of these distributions can be derived by differentiating the distribution functions. This leads to

$$\lambda(x) = \exp\{-e^{-x}\}e^{-x}, \qquad x \in \mathbb{R} \tag{2.4}$$

$$\phi_\alpha(x) = \alpha\exp\{-x^{-\alpha}\}x^{-(1+\alpha)}, \qquad x \geq 0 \tag{2.5}$$

$$\psi_\alpha(x) = \alpha\exp\{-(-x)^\alpha\}(-x)^{\alpha-1}, \qquad x \leq 0. \tag{2.6}$$

See Fig. 2.5 for examples of possible shapes of the densities. As one can see all the densities are unimodal, e.g., the density is non-decreasing left of some point u and non-increasing right of u (u is then called the mode). The Fréchet and Gumbel densities are skewed to the right. Full models usually have added some location and scale parameters μ and σ, e.g., if a random variable X has the c.d.f. F, then $\mu + \sigma X$ has the c.d.f. $F_{\mu,\sigma}(x) = F((x - \mu)/\sigma)$.

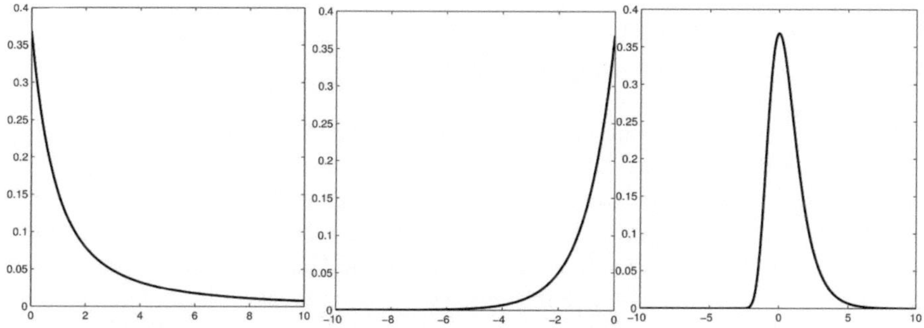

Fig. 2.5 Densities of the Fréchet, Weibull and Gumbel distribution ($\alpha = 1$)

Observe, while the distributions are usually used to model quite different phenomena from a mathematical point of view they are closely linked. For example, suppose $X > 0$, then

$$X \sim \Phi_\alpha(x) \Longleftrightarrow lnX^\alpha \sim \Lambda \Longleftrightarrow -X^{-1} \sim \Psi_\alpha$$

which can be immediately verified, e.g.,

$$\begin{aligned} P_\Phi(lnX^\alpha \leq x) &= P_\Phi(X \leq e^{x\alpha^{-1}}) \\ &= exp\{-(e^{x\alpha^{-1}})^{-\alpha}\} \\ &= exp\{-e^{-x}\} \\ &= \Lambda(x). \end{aligned}$$

Most commonly referred to in the discussions of extreme values is the Gumbel-type distribution in hydrology (Katz et al. 2002) and the Fréchet-type in financial applications (Embrechts et al. 2013). Especially for statistical estimation purposes, a one-parameter representation of the three standard cases in one family is very useful. Definition 2.1 is widely accepted in the literature as *the* standard representation.

Definition 2.1 The Generalized Extreme Value (GEV) distribution H_ξ is defined by

$$H_\xi(x) = \begin{cases} \exp\left\{-(1+\xi x)^{-1/\xi}\right\} & \text{if } \xi \neq 0 \\ \exp\{-\exp(-x)\} & \text{if } \xi = 0, \end{cases} \tag{2.7}$$

where $1 + \xi x > 0$.

As the GEV is continuous the density can be easily derived through differentiation, leading to

$$\begin{aligned} h_\xi(x) &= \exp\left\{-(1+\xi x)^{-1/\xi}\right\}(1+\xi x)^{-(1+1/\xi)} && \text{if } \xi \neq 0 \\ &= \exp\{-\exp(-x)\}\exp^{-x} && \text{if } \xi = 0. \end{aligned}$$

The unification of the three types of extreme value distributions into one single family especially simplifies statistical implementations. The parameter ξ is called the shape parameter and may be used to model a wide range of tail behavior. The type 2 (Frechet) and type 3 (Weibull) classes of extreme value distribution correspond, respectively, to the cases $\xi > 0$ and $\xi < 0$, while the Gumbel distribution arises in the limit as $\xi \to 0$. One can also introduce the related location-scale family $H_{\xi,\mu,\sigma}$ by replacing the argument x above by $(x - \mu)/\sigma$ for $\mu \in \mathbb{R}$ and $\sigma > 0$. The support has to be adjusted accordingly.

From a practical point of view Theorem 2.2 and Definition 2.1 states that if one assumes that if n (now fixed) is large enough for the limit law to be used as an approximation the GEV may be fitted directly to a series of observations of M_n.

The problem that the normalizing constants will be unknown in practice is easily resolved. Assuming Theorem 2.2 holds, e.g.,

$$P((M_n - d_n)/c_n \leq x) \approx H_\xi(x)$$

for large enough n. Then equivalently

$$P(M_n \leq x) \approx H_\xi((x - d_n)/c_n) = H_\xi^*(x),$$

where H_ξ^* is another member of the GEV family. Since the parameters of the distribution have to be estimated anyway, it is irrelevant in practice that the parameters of the distribution H are different from those of H^*. This argument will be used for estimating the parameters of the GEV distribution for block maxima data.

The next definition is very important for determining the class of nondegenerate limit distributions of maxima. It will be also used within the discussion of threshold approaches.

Definition 2.2 A nondegenerate r.v. X (or the corresponding distribution or density function) is called **max-stable** if it satisfies

$$max(X_1, .., X_n) \stackrel{d}{\rightarrow} c_n X + d_n \tag{2.8}$$

for i.i.d. $X, X_1, ..., X_n$, appropriate constants $c_n > 0$, $d_n \in \mathbb{R}$ and every $n \geq 2$.

Observe, if we assume that (X_n) is a sequence of i.i.d. max-stable r.vs. then (2.8) may be rewritten as

$$\frac{M_n - d_n}{c_n} \stackrel{d}{\rightarrow} X.$$

For illustration purposes, suppose F is again standard exponential distributed, i.e., $F(x) = 1 - exp(-x)$. To show that the maxima M_n of a sequence of i.i.d r.vs. (X_i) of this distribution is max-stable (we already have shown it converges to a Gumbel distribution but use here a slightly different notation) we choose the normalizing constants $c_n = 1$ and $d_n = ln(n)$. Then

$$\begin{aligned}
(M_n - d_n)/c_n &\stackrel{d}{\rightarrow} X\\
P((M_n - d_n)/c_n \leq x) &\longrightarrow H(x)\\
P(M_n - \ln(n) \leq x) &\longrightarrow H(x)\\
P(M_n \leq x + \ln(n)) &\longrightarrow H(x)\\
\{F(x + \ln(n))\}^n &\longrightarrow H(x)\\
\{1 - \exp(-(x + \ln(n)))\}^n &\longrightarrow H(x)\\
\{1 - (1/n)\exp(-x)\}^n &\longrightarrow \exp(-e^{-x}) \in H_\xi(x).
\end{aligned}$$

As one can see there is a connection of max-stability with the extreme limit laws which is stated in the next theorem.

Theorem 2.3 *The class of max-stable distributions coincides with the class of all possible (nondegenerate) limit laws for maxima (properly normalized) of i.i.d. r.v.*

Or in other words, a distribution is max-stable if, and only if, it is a generalized extreme value distribution (see the example of the standard exponential distribution). The theorem gives some insight into the structural stability of extreme value distributions: the distribution of maxima of i.i.d. r.v. will be of the same type as that of the underlying population iff (if and only if) the underlying population has itself a distribution of an extreme value type and they can only be one of three types (Theorem 2.2).

Another important definition in the context of extremes is the one about the maximum domain of attraction which will be also needed within the threshold approach discussion.

Definition 2.3 X belongs to the **maximum domain of attraction** of the extreme value distribution H if there exist constants $c_n > 0$, $d_n \in \mathbb{R}$ such that

$$\frac{M_n - d_n}{c_n} \xrightarrow{d} H \tag{2.9}$$

holds. We write $X \in MDA(H)$ or $F \in MDA(H)$.

Because the extreme value distributions are continuous on $\mathbb{R}$, (2.9) is equivalent to

$$\lim_{n\to\infty} P(M_n \le c_n x + d_n) = \lim_{n\to\infty} F^n(c_n x + d_n) = H(x), x \in \mathbb{R}.$$

We already indicated that quantiles correspond to the "inverse" of a distribution function and we now more generally define

Definition 2.4 The generalized inverse of the distribution function F

$$F^{-1}(t) = \inf\{x \in \mathbb{R} : F(x) \ge t\} \qquad 0 < t < 1 \tag{2.10}$$

is called the **quantile function** of the distribution F. $x_t = F^{-1}(t)$ defines the t-quantile of F.

To identify the domain of attraction of a given distribution $F(x)$ and the associated sequences c_n and d_n, the following theorem, using the quantile function, is very useful.

Theorem 2.4 *A necessary and sufficient condition for the continuous cdf F(x) to belong to the maximum domain of attraction of the distribution $H_\xi(x)$, is that*

$$\lim_{\varepsilon\to 0} \frac{F^{-1}(1-\varepsilon) - F^{-1}(1-2\varepsilon)}{F^{-1}(1-2\varepsilon) - F^{-1}(1-4\varepsilon)} = 2^{\xi}, \tag{2.11}$$

where ξ is the shape parameter of the associated limit distribution.

This implies that (i) if $\xi < 0, F(x)$ belongs to the maximum domain of attraction of the Weibull distribution (ii) if $\xi = 0, F(x)$ belongs to the Gumbel maximum domain of attraction and (iii) if $\xi > 0, F(x)$ belongs to the maximum domain of attraction of the Fréchet distribution.

For illustration, we look again at the standard exponential distribution $F = 1 - exp(-x)$, and we already know that the inverse is of the form $x_p = F^{-1}(p) = -ln(1 - p)$ and therefore

$$\lim_{\varepsilon \to 0} \frac{-ln(1-(1-\varepsilon)) + ln(1-(1-2\varepsilon))}{-ln(1-(1-2\varepsilon)) + ln(1-(1-4\varepsilon))}$$
$$\lim_{\varepsilon \to 0} \frac{-ln(\varepsilon) + ln2 + ln(\varepsilon)}{-ln2 - ln(\varepsilon) + ln4 + ln(\varepsilon)} = 1 = 2^0,$$

which shows that the standard exponential distribution belongs to the Gumbel domain of attraction for the maxima. Table 2.2 shows domains of attraction of the most common distributions used in this book.

The GEV provides a model for the distribution of block maxima, i.e., separating the data into blocks of equal length, taking the maximum from each block and using this set of data to fit a GEV. One of the most critical issues here is the selection of the block size. There is a bias-variance trade-off: too small blocks will mean that the approximation by the limit model is likely to be poor, leading to bias in the estimation and extrapolation; on the other hand, large blocks will generate only few block maxima data, leading to a large estimation variance. Most of the time diagnostic plots can be used to find a suitable compromise.

We introduced above an additional location and scale parameter for the GEV distribution which we denoted $H_{\xi,\mu,\sigma}$ and for simplicity reasons now introduce the parameter vector $\theta = (\xi, \mu, \sigma)$ and we write instead H_θ. In what follows we some-

Table 2.2 Maximum domain of attraction for various distributions

Distribution	Maximum Domain of attraction
Normal	Gumbel
Exponential	Gumbel
Lognormal	Gumbel
Gamma	Gumbel
Rayleigh	Gumbel
Uniform	Weibull
Beta	Weibull
Cauchy	Fréchet
Burr	Fréchet
Loggamma	Fréchet
Pareto	Fréchet

Source Based on Kotz and Nadarajah (2000)

times distinguish the two cases by either writing H_ξ or H_θ depending on the case at hand. Standard methodologies from parametric estimation theory like the maximum-likelihood technique are available if the given data consist of a i.i.d. sample $X_1, ..., X_n$ from H_θ. It should be noted that the assumption of X_i having an *exact* extreme value distribution H_ξ is perhaps not the most realistic one and therefore also semi-parametric approaches exist (see the discussion in Embrechts et al. 2013). However, here we limit our discussion on the main techniques for estimating θ in the exact model. In the following we assume that the sequence $X_1, ..., X_n$ is already the block-max data.

The most common technique for parameter estimation of the GEV is the maximum-likelihood method. Let's assume that H_θ has the density h_θ. The *likelihood function* which is based on the data $\mathbf{X} = (X_1, ..., X_n)$ is given by $L(\theta; \mathbf{X}) = \prod_{i=1}^n h_\theta(X_i)$ $I_{\{1+\xi(X_i-\mu)/\sigma>0\}}$. If we denote the *log-likelihood function* by $l(\theta; \mathbf{X}) = \ln L(\theta; \mathbf{X})$ the *maximum-likelihood estimator (MLE)* for θ is $\hat{\theta}_n = \arg\max_{\theta\in\Theta} l(\theta; \mathbf{X})$. Under the assumption that $X_1, ..., X_n$ are i.i.d variables, following the GEV distribution, the log-likelihood for the GEV parameters given $\xi \neq 0$ is

$$l(\xi, \mu, \sigma) = -n\log\sigma - (1 + 1/\xi) \sum_{i=1}^{n} log \left[1 + \xi\frac{x_i - \mu}{\sigma}\right] - \sum_{i=1}^{n} \left[1 + \xi\frac{x_i - \mu}{\sigma}\right]^{-1/\xi} \tag{2.12}$$

provided that $1 + \xi((x_i - \mu)/\sigma) > 0$ for $i = 1, ..., n$.

Parameter combinations, where the inequality above is violated, correspond to configurations where at least one of the data points is beyond an endpoint of the distribution. As before, the case $\xi = 0$ has to be treated separately using the Gumbel limit of the GEV distribution.

$$l(0, \mu, \sigma) = -n\log\sigma - \sum_{i=1}^{n} \left(\frac{x_i - \mu}{\sigma}\right) - \sum_{i=1}^{n} \exp\left(-\frac{x_i - \mu}{\sigma}\right). \tag{2.13}$$

Maximization of (2.12) and (2.13) with respect to the parameter vector θ leads to the maximum-likelihood estimate with respect to the entire GEV family.

Because there is no analytical solution available, some numerical optimization algorithms have to be used instead (see the discussion in Coles et al. 2001). An extension to upper order statistics is also possible, which decreases the standard error of the MLE considerably (see the subsection further down below). There are a number of non-regularity situations associated with ξ too worth to be mentioned (taken from Coles et al. 2001): when $\xi < -1$ the maximum-likelihood estimates are difficult to be obtained, when $-1 < \xi < -1/2$ MLE estimators are obtainable but there are problems in regard to the standard asymptotic properties. The good properties of MLE estimators hold if $\xi > -1/2$. A number of other techniques to estimate the parameters of the GEV exist too, including the probability weighted moments method, the elemental percentile method, the quantile least squares method, and the truncation method. Algorithms and explanations for each technique can be found in Castillo et al. (2005). Subject to limitations already discussed in regard to the ξ parameter, the approximate distribution of $(\hat{\xi}, \hat{\mu}, \hat{\sigma})$ is multivariate normal with mean (ξ, μ, σ). Using the inverse of the observed Fisher information matrix evaluated at the maximum-likelihood estimates one can also obtain the variance-covariance matrix. Therefore, it is relatively easy to calculate confidence regions due to the approximate normality of the estimators (see again Castillo et al. 2005), e.g., the $(1 - \alpha)$ confidence interval using the respective standard errors $\hat{\sigma}$ (e.g., taking the square root from the diagonals of the covariance matrix) (see again the related discussions in Coles et al. 2001):

$$\begin{aligned} &\hat{\xi} \pm z_{\alpha/2}\hat{\sigma}_{\hat{\xi}} \\ &\hat{\mu} \pm z_{\alpha/2}\hat{\sigma}_{\hat{\mu}} \\ &\hat{\sigma} \pm z_{\alpha/2}\hat{\sigma}_{\hat{\sigma}}, \end{aligned}$$

where $z_{\alpha/2}$ is the (1-α/2) quantile of the standard normal distribution. Observe that extreme quantile estimates can be obtained by inverting the equations from Definition (2.1) to get

$$x_p = \begin{cases} \mu - \frac{\sigma}{\xi}[1 - \{-\log(1-p)\}^{-\xi}], & \xi \neq 0, \\ \mu - \sigma\log\{-\log(1-p)\}, & \xi = 0, \end{cases} \tag{2.14}$$

where $H_\theta(x_p) =$ 1-p. Usually, x_p is called the *return level* associated with the *return period* $1/p$. In other words, in the case of yearly blocks, the return level x_p is expected to be exceeded on average once every $1/p$ years or stated differently x_p is exceeded by the annual maximum in any particular year with probability p. By substituting the MLE of the GEV into (2.14) the return level x_p can easily be determined. As before, confidence regions can be calculated due to the approximate normality of the estimators. However, better approximations can be obtained from the appropriate profile likelihood function (see the discussion below).

Once the parameters for the GEV are estimated the model validity needs to be checked. Most of the time goodness of fit plots are used that compare data from the empirical and fitted distribution (the discussion is again based on Coles et al. 2001). Assume for the moment that one has ordered block maxima data denoted as $x_{(1)}, \ldots, x_{(n)}$ and define the empirical distribution function at $x_{(i)}$ as

$$\hat{F}(x_{(i)}) = i/(n+1). \tag{2.15}$$

Substituting the parameter estimate into Definition (2.1), one can get the corresponding model-based estimate $H_{\hat{\theta}}(x_{(i)})$. The *probability plot*, also called P-P plot, consists of the points

$$\{(H_{\hat{\theta}}(x_{(i)}), \hat{F}(x_{(i)})), i = 1, \ldots, n\}$$

and for a valid model the points should be close to the unit diagonal. A substantial departure from linearity usually indicates some failure or problems in the GEV model estimation. The main disadvantage of a P-P plot is that it provides the least information in the region of most interest, e.g., accuracy of the model for large values of x. To overcome this limitation, a more common plot used is the so-called *quantile-quantile plot* or Q-Q plot which consists of the points

$$\{(H_{\hat{\theta}}^{-1}(i/(n+1)), (x_{(i)})), i = 1, ..., n\}$$

with

$$H_{\hat{\theta}}^{-1}\left(\frac{i}{n+1}\right) = \hat{\mu} - \frac{\hat{\sigma}}{\hat{\xi}}\left[1 - \left\{-log\left(\frac{i}{n+1}\right)\right\}^{-\hat{\xi}}\right].$$

Hence, the Q-Q plot shows the estimated versus the observed quantiles. If the model fit is good the pattern of points will follow a 45° straight line. Again, departures from linearity in the quantile plot indicate some model failure. Observe, if the data is plotted against the exponential distribution concave departures from a straight line indicate a heavy-tailed distribution, whereas convex departures can be interpreted as thin tail behavior.

Another very useful so-called *return level plot* consists of the locus of points

$$\left\{(ln(-ln(1-p)), \hat{x}_p) : 0 < p < 1\right\},$$

where $\hat{x}_p$ is the estimated x_p from the GEV model. Also here departures from the straight line indicate problems with the estimated model. Last but not least, one important plot is based on the *profile log-likelihood function*

$$l_p(\theta_i) = max_{\theta_{-i}} l(\theta_i, \theta_{-i}),$$

where θ_{-i} denotes all components of θ excluding θ_i. To obtain a profile log-likelihood for ξ, one needs to fix ξ_0 and maximize the log-likelihood with respect to the other parameters, e.g., μ and σ. This operation is done for a range of ξ_0. The corresponding maximized values of the log-likelihood constitute the profile log-likelihood for ξ. The same procedure can be applied for the return levels too. Profile log-likelihood based inferences often give a more accurate representation of uncertainty. For example, the skewness of the profile may indicate greater uncertainty in the higher return levels.

We want to use the already introduced weather data from Malawi to estimate extreme daily precipitation events for the rainy season. This reduces the sample size already considerably to only 44 data points. Applying the maximum-likelihood technique the estimated parameters are $(\hat{\xi}, \hat{\mu}, \hat{\sigma}) = (0.09, 62.03, 14.53)$ which indicates a Frechet distribution, i.e., a heavy tail. The diagnostic plots as discussed above are shown in Fig. 2.6 and indicate a reasonable fit.

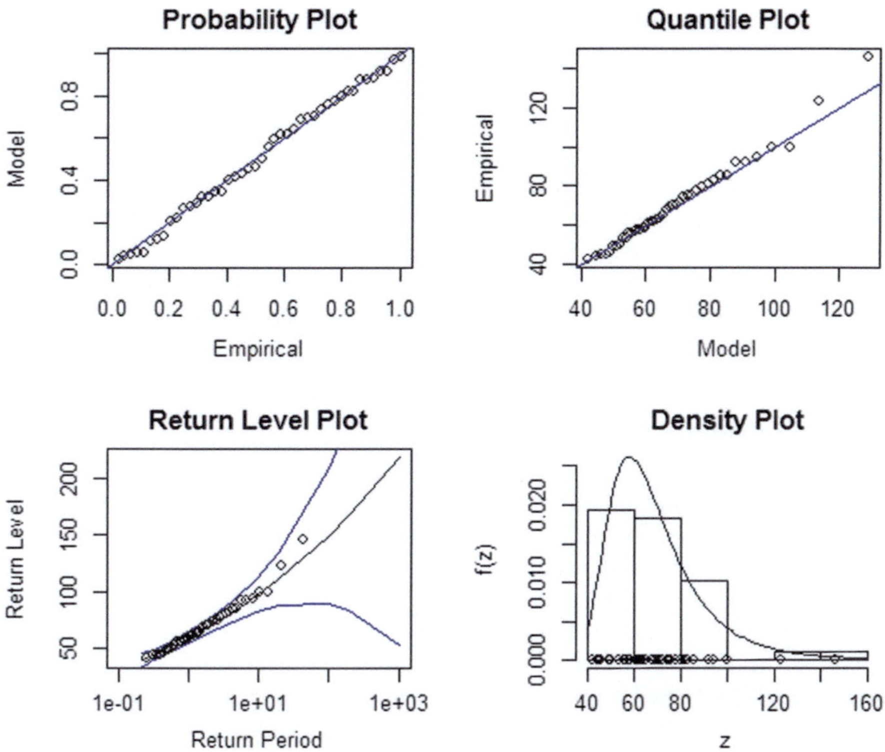

Fig. 2.6 Diagnostic plots of extreme daily rainfall

Also confidence regions can be calculated using the covariance matrix as discussed above and quite some large uncertainties around the estimates exist. Especially the 95% confidence interval for the shape parameter includes zero and we therefore rather assume a Gumbel distribution which also shows good diagnostic plots; the related location and scale parameters are estimated to be 62.77 and 15.12. Assuming a correct model with the estimates given above one can now determine various return levels and return periods. For example, the probability that maximum daily rainfall is below 100 mm within the rainy season can be calculated as $P(X \leq 100) = exp(-exp(-\frac{100-62.77}{15.12})) = 0.92$. In other words, given our Gumbel distribution assumption, the probability that the maximum daily rainfall over the rainy season will be greater than 100 is 8% or will occur, on average, every 12 (=1/0.08) years. This is a conditional probability, i.e., it is the probability that rainfall is greater than 100 given that the maximum occurs. We do not know at what time in the season the maximum occurs but when it does the probability that it is greater than 100 is 8%. The unconditional probability $P(X > 100)$ can be calculated as $P(X > 100) = P(X > 100|max)P(max) + P(X > 100|notmax)P(notmax)$. In our example, a 182-day time window is used (October to March), hence, the probability that any given day is a maximum is 1/182. Therefore, $P(X > 100) =$

$0.08 * \frac{1}{182} + P(X > 100|notmax) * \frac{181}{182}$. $P(X > 100|notmax)$ are the non-extreme events and can be estimated through another parametric distribution as discussed in Sect. 2.1.

As was indicated in the beginning, the assumption of $\{X_i\}_{i=1}^n$ following an exact extreme value distribution H_ξ is perhaps not always realistic. However, one can assume that the X_i are *approximately* H_ξ distributed, e.g., approximately will be interpreted as belonging to the "maximum domain of attraction of" and in this case different estimators such as Pickands Estimator, Hills Estimator, or the Deckers-Einmahl-de Haan Estimator can be used. We refer to Embrechts et al. (2013) for an introduction and we focus now on threshold modeling approaches.

2.2.2 *Distribution of Exceedances*

There are some problems when a block maxima approach is applied. First of all, one has to choose the time period defining the blocks where the maximum is chosen from, e.g., days, months, or even years. As a consequence usually the total sample size is very much reduced and estimations are based on only few data points. To avoid such problems one can also take a look at maxima data points which are defined to be above some pre-defined threshold level, e.g., only data points are looked at which exceed a given threshold level, say u (see Fig. 2.4). Also in this case extreme value theory provides a powerful result, similar to the GEV, but now about the distribution function over a given threshold level. For simplicity reasons, we define $\overline{F}$ as $\overline{F} = 1 - F$. The following theorem is needed to proceed forward and the following discussion is mostly based on Embrechts et al. (2013). We already introduced the Maximum Domain of Attraction which will be used in the following theorem.

Theorem 2.5 *For $\xi \in \mathbb{R}$ the following assertions are equivalent:*

$$(i) F \in MDA(H_\xi) \tag{2.16}$$

(ii) There exists a positive, measurable function $a(\cdot)$ such that for $1 + \xi x > 0$,

$$lim_{u \uparrow x_F} \frac{\overline{F}(u + xa(u))}{\overline{F}(u)} = \begin{cases} (1+\xi x)^{-1/\xi} & if\, \xi \neq 0 \\ e^{-x} & if\, \xi = 0. \end{cases} \tag{2.17}$$

The condition (2.17) can be reformulated as follows. Let X be a r.v. with $F \in MDA(H_\xi)$, then

$$lim_{u \uparrow x_F} P\left(\frac{X-u}{a(u)} > x|X > u\right) = \begin{cases} (1+\xi x)^{-1/\xi} & \text{if } \xi \neq 0 \\ e^{-x} & \text{if } \xi = 0 \end{cases}$$

which gives a distributional approximation for scaled excesses over a high threshold level u. This gives rise to the following definition.

Definition 2.5 The Generalized Pareto distribution (GPD) G_ξ is defined as

$$G_\xi(x) = \begin{cases} 1 - (1+\xi x)^{-1/\xi} & \text{if } \xi \neq 0 \\ 1 - e^{-x} & \text{if } \xi = 0, \end{cases} \tag{2.18}$$

where $x \geq 0$ if $\xi \geq 0$ and $0 \leq x \leq -1/\xi$ if $\xi < 0$.

As in the GEV case also here one can introduce the related location-scale family $G_{\xi,\mu,\sigma}$ by replacing the argument x by $(x-\mu)/\sigma$ for $\mu \in \mathbb{R}$ and $\sigma > 0$. Important here is the case where $\mu = 0$ and for notational convenience we will denote it by $G_{\xi,\sigma}$:

$$G_{\xi,\sigma}(x) = \begin{cases} 1 - (1+\xi \frac{x}{\sigma})^{-1/\xi} & \text{if } \xi \neq 0 \\ 1 - e^{-x/\sigma} & \text{if } \xi = 0, \end{cases}$$

where the domain $D(\xi, \sigma) = [0, \infty]$ for $\xi \geq 0$ and $[0, -\sigma/\xi]$ if $\xi < 0$. As we are dealing with excesses, we define next the excess distribution function. To avoid confusion we denote the excesses over u by $Y_1, ..., Y_n$.

Definition 2.6 Let X be a r.v. with c.d.f. F and right endpoint x_F. Then,

$$F_u(y) = P(X - u \leq y \mid X > u), \text{ where } 0 \leq y \leq x_F - u \tag{2.19}$$

is the **excess distribution function** of the r.v. X over the threshold u. The function

$$e(u) = E(X - u \mid X > u) \tag{2.20}$$

is called the **mean excess function**.

As in the block maxima case also here we are dealing with conditional probabilities, in this case in the sense that a given loss is smaller or equal than $x - u$ *given* that a maximum u is exceed (see Fig. 2.7). This is not the same as the unconditional probability which is simply the probability that a random variable is below a given value. In that regard, observe that F_u can be written in terms of F,

$$F_u(y) = \frac{P(X - u \leq y) \cap P(X > u)}{P(X > u)} = \frac{F(y+u) - F(u)}{1 - F(u)}. \tag{2.21}$$

Therefore, the unconditional probability can be written as $F(x) = (1 - F(u))$ $F_u(y) + F(u)$. Note, $F(u)$ can be estimated by $(n - n_u)/n$ ($n =$ total number of observations and $n_u =$ total numbers of exceedances) while $F_u(y)$ can be approximated as discussed below.

Restating our Definition 2.5 before will give the following final *classic* result for the threshold approach:

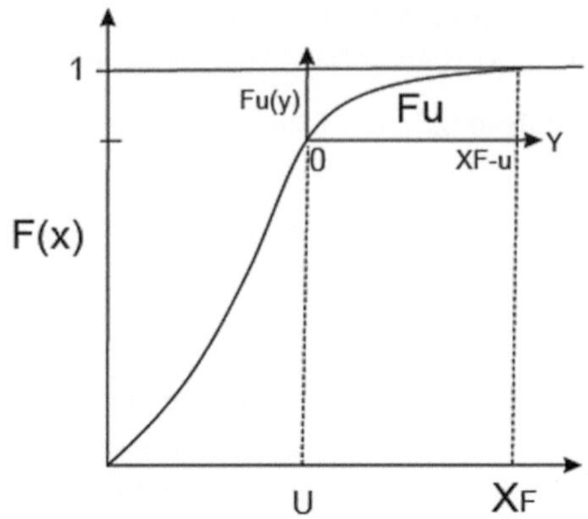

Fig. 2.7 Distribution function F and conditional distribution function F_u

Definition 2.7 For a large class of distribution functions F (i.e., $F \in MDA(H)$), the excess distribution $F_u(y)$ can be approximated for large u ($u \to \infty$), by

$$F_u(y) \approx G_{\xi,\sigma}(y), \tag{2.22}$$

where

$$G_{\xi,\sigma}(y) = \begin{cases} 1 - (1 + \frac{\xi y}{\sigma})^{-1/\xi} & \text{if } \xi \neq 0 \\ 1 - e^{-y/\sigma} & \text{if } \xi = 0 \end{cases} \tag{2.23}$$

for $y \in [0, (x_F - u)]$ if $\xi \geq 0$ and $y \in [0, -\sigma/\xi]$ if $\xi < 0$. $G_{\xi,\sigma}(x)$ is called the Generalized Pareto Distribution (GPD).

We can therefore summarize. The generalized extreme value distribution H_θ with $\xi \in \mathbb{R}$, $\mu \in \mathbb{R}$, $\sigma > 0$ describes limit distributions of normalized maxima. The generalized Pareto distribution $G_{\xi,\sigma}$, with $\xi \in \mathbb{R}$, $\sigma > 0$ appears as the limit distribution of scaled excesses over high thresholds (Embrechts et al. 2013). One important task is however to determine the threshold level so that the theorem approximately holds. The issue of threshold selection is similar to the choice of the block size in the block maxima approach as there is again a bias-variance trade-off. Too low thresholds lead to the violation of the asymptotic basis of the model, too high thresholds lead to fewer data points and higher variances. We present here first an exploratory technique mostly used prior to the model estimation.

Suppose the data consists of i.i.d. measurements and let $x_{(1)}, \ldots, x_{(n_u)}$ be the subset of data points that exceed a particular threshold u. The *mean residual life plot* is the locus of points:

$$\left\{ \left(u, \frac{1}{n_u} \sum_{i=1}^{n_u} (x_{(i)} - u) \right) : u < x_{max} \right\}, \tag{2.24}$$

where n_u is the number of observations above the threshold u. Above a threshold which provides a valid approximation to the excess distribution, the mean residual life plot should be approximately linear in u. It should be noted that threshold selection is actually not an easy task due to the bias-variance trade-off and the interpretation of a mean residual life plot is not always simple in practice. Nevertheless, it is one of the most practical tools for threshold selection.

In the case of our daily rainfall data during the rainy season in Malawi we use now all data points for our analysis of a possible GPD. It should be noted that the block maxima approach is usually used when seasonality occurs and in the case of thresholds also wet or dry periods should be distinguished. Nevertheless, we use here the full season for our analysis. Figure 2.8 shows the mean residual life plot.

After having determined the threshold, the parameters of the GPD distribution can be estimated, in a similar fashion as in the GEV case, e.g., via maximum likelihood. Assume that the values $y_1, ..., y_{n_u}$ are the n_u excesses over a threshold u, e.g., $y_j = x_{(j)} - u$ for $j = 1, ..., n_u$. For $\xi \neq 0$ the log-likelihood is

$$l(\xi, \sigma) = -n_u \log \sigma - (1 + 1/\xi) \sum_{i=1}^{n_u} \log(1 + \xi y_i/\sigma)$$

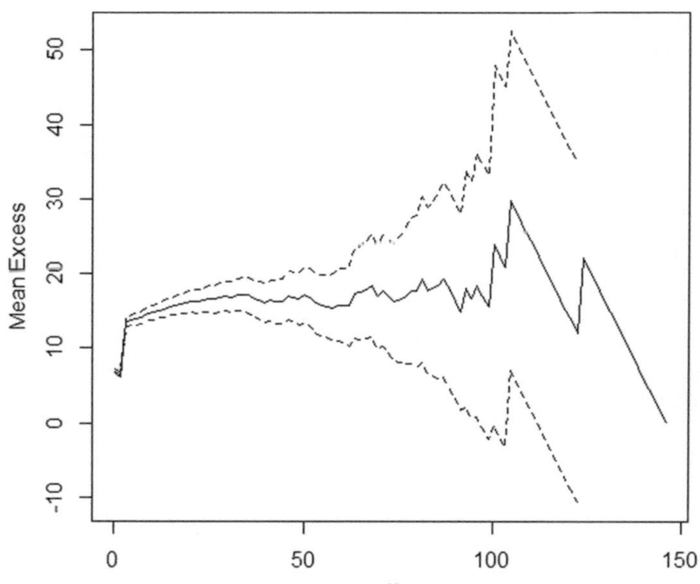

Fig. 2.8 Mean residual life plot for daily rainfall in Chitedze

provided that $(1 + \xi y_i/\sigma) > 0$ for $i = 1, ..., n_u$; otherwise $l(\xi, \sigma) = -\infty$. For the case $\xi = 0$ the log-likelihood is

$$l(\sigma) = -n_u \log \sigma - \frac{1}{\sigma} \sum_{i=1}^{n_u} y_i.$$

As in the GEV case also here analytical results are not available and numerical techniques have to be used instead. Especially for $\xi > -1/2$ the ML method works fine. Also similarly to GEV, model checking can be done again graphically either using probability plots, quantile plots, return level plots, or through density plots. Going back to our Malawi example and using the mean excess plot presented above we determine as a first threshold level 40 which gives 172 observations to be used for the estimation (e.g., nearly four times more than in the block maxima case). Also here one can perform some diagnostic model checking. Figure 2.9 shows the log-likelihood plot for the shape parameter. As in the GEV case also here the shape parameter should be assumed to be zero. The scale parameter is estimated to be around 16.4. We can again ask the question about the probability that a 100 rainfall amount is exceeded given that we are above a 40 rainfall level. One has $100 - 40 = 60$ and therefore $P(X > 100) = 1 - P(X \leq 100) = 1 - (1 - exp(-60/16.4) = 0.026$. Hence, the probability that there is higher daily rainfall than 100 given that it is above 40 is 2.6%. Also here this is a conditional probability. The unconditional probability that the rainfall is above 100 can be sim-

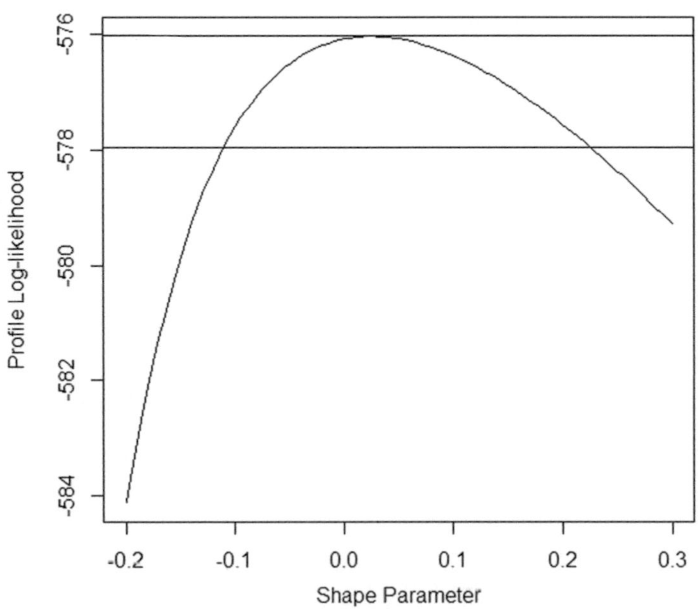

Fig. 2.9 Profile Log-likelihood of the shape parameter

ply calculated as discussed above. Additionally, one can combine the GPD for the analysis of extremes and a parametric family for the non-extreme cases. As we will see in the next section, threshold approaches naturally lead to a point process characterization of extremes.

2.2.3 *Point Process Characterization of Extremes*

It is very convenient (at least statistically) to characterize extreme value behavior from a point process perspective. Related techniques can give important insights into the structure of limit variables and limit processes, e.g., probability of the occurrence of a given number of events or the probability of the time between two successive events. The so-called Poisson process, defined further down below, plays a central role within this approach. The basic idea is to combine the occurrence of exceedances and the excesses over high thresholds jointly within a 2-dimensional non-homogenous Poisson process, i.e., one dimension is time and the other is the excess value. In this way, it is also possible to combine the features of the GEV distribution for block maxima and the POT (peak over threshold) approach for threshold exceedances, i.e., the GEV distribution can be indirectly fitted via the POT method, but still in terms of the GEV parametrization. We start with an introduction to the so-called Poisson process. Let's call the number of events (or points) happening within a finite interval of length $t > 0$ as $N(t)$. If the number of events in a given interval is following the Poisson distribution with the mean proportional to the length of the interval, and additionally the number of events occurring in disjoint intervals is mutually independent, then we call such a process a homogenous Poisson process. More formally we have the following definition.

Definition 2.8 A point process is said to be a **homogeneous Poisson process** with intensity $\lambda > 0$ if **(i)** the number of events (arrivals) $N(t)$ in a finite interval of length $t > 0$ obeys the Poisson (λt) distribution, i.e.,

$$P(N(t) = n) = \frac{(\lambda t)^n}{n!} e^{-\lambda t}, \qquad n = 1, 2, 3, \ldots \tag{2.25}$$

and **(ii)** the number of events $N(t_1, t_2)$ and $N(t_3, t_4)$ in non-overlapping intervals $t_1 \leq t_2 \leq t_3 \leq t_4$ are independent random variables.

Equivalent definitions of a Poisson process focus, for example, on pure birth processes (e.g., in an infinitesimal time interval dt there may occur only one arrival; the probability of this event is equal to λdt and is independent of arrivals outside the interval) or on the interarrival times (the interarrival times are assumed to be independent and follow the Exp(λ) distribution, i.e., P(interarrival time $> t$) $= e^{-\lambda t}$).

The so-called **non-homogeneous Poisson process** is defined as in Definition 2.8 but now the intensity is allowed to vary in time t which is usually indicated through $\lambda(t)$.

Definition 2.9 A point process is said to be a **non-homogeneous Poisson process** with intensity measure Λ (and intensity function λ) if **(i)** the number of events $N(A)$ occurring in A is a Poisson r.v., that is,

$$N(A) \sim Poi(\Lambda(A)) \text{ where } \Lambda(A) = \int_{t_1}^{t_2} \lambda(t)dt, \tag{2.26}$$

and **(ii)** for any $m > 1$, if $A_1, ..., A_m$ are mutually disjoint subsets the number of events in $N(A_1), ..., N(A_m)$ are independent r.v.

Given these definitions let's assume now that $X_1, ..., X_n$ is an i.i.d. series from an unknown distribution F. Furthermore, assume that F is in the domain of attraction of a GEV distribution and that the required constants for normalization are again the two parameters notated as c_n and d_n. As already said, the basic idea is to form a 2-dimensional point process $\{(i, X_i) : i = 1, ..., n\}$ and to characterize the behavior of this process in regions of the form $[t_1,t_2]$x(u,∞), thus giving a representation for the behavior of the X_i at large levels (here determined through the threshold u).

Similar to the block maxima and threshold approach one first constructs a sequence, in this case of point processes $P_1, P_2, ...$ on [0,1]x$\mathbb{R}$ with P_n defined as

$$P_n = \left\{\left(\frac{i}{n+1}, \frac{X_i - d_n}{c_n}\right) : i = 1, ..., n\right\} \tag{2.27}$$

and examines the limiting behavior. Observe, because we assume that F is in the domain of attraction of a GEV distribution $(M_n - d_n)/c_n$ (which is the largest point on the y-axis) is nondegenerate and therefore the limit process of P_n is also nondegenerate. Furthermore, small points are normalized (e.g., scaled toward the lower endpoint) to the same value d_l $(= \lim_{n\to\infty}(x_F - d_n)/c_n)$ and large points are retained in the limit process. It can be shown that the process P_n converges weakly to a Poisson process on sets which exclude a lower boundary. Hence, at high levels $\{(t_1, t_2) \times (x, \infty)\}$, the process P_n should approximate a Poisson process with intensity measure

$$\Lambda\{(t_1, t_2) \times (x, \infty)\} = (t_2 - t_1)[1 + \xi x]^{-1/\xi}$$

One can absorb the rescaling factors d_n and c_n into the distribution and $P_n = \left\{\left(\frac{i}{n+1}, X_i\right) : i = 1, ..., n\right\}$ can be approximated above high thresholds by a Poisson process with intensity function

$$\Lambda\{(t_1, t_2) \times (x, \infty)\} = (t_2 - t_1)[1 + \xi \left(\frac{x-\mu}{\sigma}\right)]^{-1/\xi}.$$

Hence, the additional location and scaling parameters μ and σ assume the role of the unknown normalizing constants (similar to the GEV and GPD discussion). As in the previous cases, this model can be fitted to observed data by maximum-likelihood techniques to determine the three parameters.

As already said the Poisson process model is a very general model. Both the GEV model for maxima and the GPD model for excesses can be derived from this model. To derive the GEV for block maxima from the Poisson process model assume a Poisson process for $P_n = \left\{\left(\frac{i}{n+1}, X_i\right) : i = 1, ..., n\right\}$ and consider $M_n = \max(X_1, ..., X_n)$ is less than some value $x \geq u$:

$$\begin{aligned} P(M_n \leq x) &= P\{\text{no points in}\{(0, 1] \times (x, \infty)\}\} \\ &= \exp(-\Lambda\{(0, 1] \times (x, \infty)\}) \\ &= \exp(-[1 + \xi \left(\frac{x-\mu}{\sigma}\right)]^{-1/\xi}) \\ &= H_\xi((x-\mu)/\sigma)). \end{aligned}$$

To derive a GPD from the Poisson process model consider again $P_n = \left\{\left(\frac{i}{n+1}, X_i\right) : i = 1, ..., n\right\}$ and look at the excess distribution:

$$\begin{aligned} F_u(x-u) &= 1 - P(X > x)/P(X > u) \\ &= 1 - \frac{\Lambda\{(0, 1] \times (x, \infty)\}}{\Lambda\{(0, 1] \times (u, \infty)\}} \\ &= 1 - \left(\frac{1 + \xi(x-\mu)/\sigma}{1 + \xi(u-\mu)/\sigma}\right)^{-1/\xi} \\ &= 1 - (1 + \xi(x-u)/\beta)^{-1/\xi} \\ &= G_{\xi,\beta}(x-u) \text{ with } \beta = \sigma + \xi(u-\mu). \end{aligned}$$

The advantages of a point process approach include the very general approach to extremes, e.g., GEV and GPD are implicit in this formulation. However, the estimates should be approximately constant over different thresholds u for assuming a correct Poisson process model, or else the intensity measure must be generalized to allow time-dependent parameters, e.g., to model trends. Another possible perspective on extremes which will be important for the copula approaches in Chap. 3 are the so-called k largest order statistics, discussed next.

2.2.4 *Modeling the K Largest Order Statistics*

Due to the limited amount of data when using the block maxima method different approaches like threshold methods as well as models for the k largest order statistics were created. The k largest order statistic model gives parameters which correspond to those for the GEV distribution of block maxima, but incorporates more of the observed extreme data. The basic idea here is to use not the maximum but the k largest observations in each block. So the interpretation is unaltered, but precision should be improved due to the inclusion of extra information. Nevertheless, one has to be careful that still only extremes are incorporated in the estimation procedure so that the approximation holds.

Suppose that $X_1, X_2, ...$ are i.i.d. r.v. and the aim is to characterize the extremal behavior of the X_i. We already know that the limiting distribution (suitably rescaled) as n tends to infinity of M_n is a GEV distribution (or no distribution at all) and we want to extend this result now to order statistics by defining first what we understand under the k largest of a sequence of random variables:

$$M_n^{(k)} = k\text{th largest of } X_1, ..., X_n.$$

Next, we want to discuss the limiting behavior of $M_n^{(k)}$ with Theorem 2.6.

Theorem 2.6 *If there exist sequences of constants* $\{c_n\}$ *and* $\{d_n\}$ *such that* $P\{(M_n - d_n)/c_n \le x\} \to H_\xi(x)$ *as* $n \to \infty$ *for some nondegenerate distribution function* H, *so that* H *is the GEV, then for fixed* k, $P\{(M_n^{(k)} - d_n)/c_n \le x\} \to H_\xi^k(x)$ *on* $\{x : 1 + \xi(x - \mu)/\sigma > 0\}$, *where*

$$H_\xi^k(x) = exp\{-\tau(x)\} \sum_{s=0}^{k-1} \frac{\tau(x)^s}{s!} \tag{2.28}$$

$$\text{with } \tau(x) = \left[1 + \xi\left(\frac{x-\mu}{\sigma}\right)\right]^{-1/\xi}. \tag{2.29}$$

So surprisingly, if the kth largest in a block is normalized in exactly the same way as the maximum, then its limiting distribution is of the form given by 2.28, the parameters of which correspond to the parameters of the limiting GEV distribution of the block maxima.

While Theorem 2.6 gives a family for the approximate distribution of each of the components of $\mathbf{M}_n^{(k)} = (M_n^{(1)}, ..., M_n^{(k)})$, it does not give the joint distribution of $\mathbf{M}_n^{(k)}$. An additional difficulty is that the components can not be independent, so the outcome of each component influences the other. Nevertheless, the following theorem gives the joint density function of the limit distribution which is very useful for estimation purposes

Theorem 2.7 *If there exist sequences of constants* $\{c_n\}$ *and* $\{d_n\}$ *such that* $P\{(M_n - d_n)/c_n \le x\} \to H_\xi(x)$ *as* $n \to \infty$ *for some nondegenerate distribution function* H, *then for fixed* k, *the limiting distribution as* $n \to \infty$ *of*

$$\hat{\mathbf{M}}_n^{(k)} = \left(\frac{M_n^{(1)} - d_n}{c_n}, ..., \frac{M_n^{(k)} - d_n}{c_n}\right)$$

falls within the family having joint probability density function

$$f(x^{(1)},...,x^{(k)}) = exp\left\{-\left[1+\xi\left(\frac{x^{(k)}-\mu}{\sigma}\right)\right]^{-1/\xi}\right\}$$
$$\times\prod_{i=1}^{k}\sigma^{-1}\left[1+\xi\left(\frac{x^{(i)}-\mu}{\sigma}\right)\right]^{-\frac{1}{\xi}-1},$$

where $-\infty < \mu < \infty, \sigma > 0$*, and* $-\infty < \xi < \infty$*;* $x^{(k)} \leq x^{(k-1)} \leq ... \leq x^{(1)}$*; and* $x^{(i)} : 1 + \xi(x^{(i)} - \mu)/\sigma > 0$ *for* $i = 1, \ldots, k$*.*

In the case of $k = 1$, Theorem 2.7 is reduced to the GEV family or density function. The case $\xi = 0$ is as before interpreted as the limiting form as $\xi \to 0$ leading to the family of density functions

$$f(x^{(1)},...,x^{(k)}) = \exp\left\{-\exp\left[-\frac{x^{(k)}-\mu}{\sigma}\right]\right\}$$
$$\times\prod_{i=1}^{k}\sigma^{-1}\exp\left[-\frac{x^{(i)}-\mu}{\sigma}\right]$$

and also here for the case $k = 1$ the formula reduces to the density of the Gumbel family.

Summarizing, similar to the block maxima technique we assume for the k largest order model that the data of i.i.d. r.v. are grouped into m blocks and in each block i the largest k_i observations are recorded. This leads to a sequence $\mathbf{M}_i^{(k_i)} = (x_i^{(1)}, ..., x_i^{(k_i)})$, for $i = 1, ..., m$. Using the density functions presented above, one can obtain numerically the ML estimates. Standard asymptotic likelihood theory also gives approximate standard errors and confidence intervals. In the special case of $k_i = 1$ for each i the likelihood function reduces the likelihood of the GEV model for block maxima.

2.2.5 *Temporal Dependence and Non-stationarity Issues*

The approaches and techniques presented so far have assumed a series of independent and identical distributed random variables. This is clearly an oversimplification and in practice one has to account for violation of these assumptions, especially in regards to stationarity as well as dependency dimensions. We therefore discuss now in some detail temporal dependence and non-stationarity issues frequently encountered in applied analysis of extremes. The name *temporal* dependence already indicates that the observations are usually obtained one after the other as time progresses. In the context of extremes, two sorts of temporal dependence are of special interest, namely, (i) short-term dependence and (ii) long-run dependence. In extreme event applications, long-range dependence is usually much less likely on physical grounds (for example, heavy rainfall today might influence the probability of extreme rainfall in the next day but not for a specified day in, say, half a year) than short-term dependence, for example, storms tend to cluster in time. In both cases the important question is how such dependence may affect the asymptotics for extreme value analysis, e.g., can the results derived in the previous sections also applied to cases where time dependency can be observed. The condition that is generally used to answer related questions is the so-called $D(u_n)$ condition (Leadbetter et al., 1983). The discussion follows again Coles et al. (2001). Recall first that a random sequence $\{X_i\}_{i=1}^{\infty}$ is called strictly stationary if $\{X_{n+k}\}_{n=1}^{\infty}$ has the same distribution as $\{X_n\}_{n=1}^{\infty}$ for all positive integers k. Then

Definition 2.10 A stationary series $X_1, X_2, ...$ is said to satisfy the $D(u_n)$ condition if, for all $i_1 < ... < i_p < j_1 < ... < j_p$ with $j_1 - i_p > l$,

$$| F_{i_1,...i_p,j_1,...,j_p}(u_n) - F_{i_1,...i_p}(u_n)F_{j_1,...,j_p}(u_n) | \leq \alpha(n, l), \tag{2.30}$$

where $\alpha(n, l_n) \to 0$ for some sequence l_n such that $l_n/n \to 0$ as $n \to \infty$ (or $l_n = o(n)$).

One can immediately observe that for sequences of independent variables in (2.30), the difference in probabilities is exactly zero for *any* sequence u_n (as independence means that for example $P(X_1 \leq u, X_2 \leq u) = P(X_1 \leq u)P(X_2 \leq u)$). The condition can be used to show that weak long-range dependence is negligible in regard to the asymptotic limit which is stated more formally in the following theorem.

Theorem 2.8 *Let $X_1, X_2, ..., X_n$ be a stationary process and define $M_n = max\{X_1, ..., X_n\}$. If $\{c_n\}$ and $\{d_n\}$ are sequences of constants such that*

$$P((M_n - d_n)/c_n \leq x) \to H(x), \tag{2.31}$$

where H is a nondegenerate distribution function, and the $D(u_n)$ condition is satisfied with $u_n = c_n x + d_n$ for every real x, H is a member of the GEV family of distributions.

Quite informally we can summarize that the maximum of a stationary sequence that has some dependence at long range follows a GEV distribution. Unfortunately, in contrast to weak long-range dependence, short-term dependence has some effect on the asymptotic limits. For illustration, we discuss this issue with an example taken from Coles et al. (2001) (Example 5.1, p. 94) that will conveniently lead to one important theorem of how to deal with short-term dependency.

Assume that $Y_0, Y_1, Y_2, ...$ is a sequence of i.i.d r.v. with distribution function $F_Y(x) = \exp(-\frac{1}{(a+1)y})$, $y > 0$, $0 \leq a \leq 1$. Define another random variable X_i by $X_0 = Y_0$ and $X_i = \max(aY_{i-1}, Y_i)$ for $i = 1, ..., n$. It follows that $P(X_i \leq x) = P(aY_{i-1} \leq x, Y_i \leq x) = P(aY_{i-1} \leq x)P(Y_1 \leq x) = \exp(-1/x)$ given $x > 0$. Therefore, the marginal distribution of the X_i series for $i = 1, ..., n$ is standard Fréchet. Furthermore, this series is stationary. Now, let $X_1^*, X_2^*, ...$ be a series of i.i.d. r.v with marginal standard Fréchet distribution, and define $M_n^* = \max(X_1^*, ..., X_n^*)$.

The comparison of the limit distribution of both maxima will show how dependency affects the behavior of the asymptotic result. First, $P(M_n^* \leq nx) = [\exp(-1/(nx))]^n = \exp(-1/x)$. On the other hand, for $M_n = \max(X_1, ..., X_n)$,

$$\begin{aligned} P(M_n \leq nx) &= P(X_1 \leq nx, ..., X_n \leq nx) \\ &= P(Y_1 \leq nx, aY_1 \leq nx, ..., aY_{n-1} \leq nx, Y_n \leq nx) \\ &= P(Y_1 \leq nx, ..., Y_n \leq nx) \qquad \text{because } 0 \leq a \leq 1 \\ &= \left[\exp\left(-\frac{1}{(a+1)nx}\right)\right]^n = (\exp(-1/x))^{\frac{1}{a+1}}, \end{aligned}$$

It follows that $P(M_n^* \leq nx) = (P(M_n \leq nx))^{1/(a+1)}$. Observe, while the marginal distribution of each series is the same, with increasing values of a, there is a tendency for extreme values to occur in groups. For example, with $a = 1$ the exponent in the last expression above is 1/2 which is the reciprocal of the cluster size and also the maximum has a tendency to decrease as a increases. Summarizing, while we obtain a limiting distribution for both sequences and they are of the same type they are also linked through the exponent. Based on this example the next theorem shows the exact linkage between the distributions which holds for a wide class of stationary processes under suitable regular conditions.

Theorem 2.9 *Let $X_1, X_2, ...$ be a stationary process and $X_1^*, X_2^*, ...$ be a sequence of independent variables with the same marginal distribution. Define $M_n = max(X_1, ..., X_n)$ and $M_n^* = max(X_1^*, ..., X_n^*)$. Under the condition that the stationary process satisfies $D(u_n)$ and (the so-called extremal index) $\theta > 0$ exists*

$$P((M_n^* - d_n)/c_n \leq x) \to G_1(x) \tag{2.32}$$

as $n \to \infty$ for normalizing sequences $c_n > 0$ and d_n, where G_1 is a nondegenerate distribution function, iff

$$P((M_n - d_n)/c_n \leq x) \to G_2(x), \tag{2.33}$$

where

$$G_2(x) = G_1^{\theta}(x) \tag{2.34}$$

for a constant θ such that $0 < \theta \leq 1$.

Informally this means that the asymptotic distribution of maxima with processes having short-range dependence characterized by the so-called extremal index θ is also a GEV distribution (the discussion of the extremal index will be done in Chap. 3). Moreover, the only effect of dependence is a powering of the limit distribution by θ which in turn only affects the location and scale parameters σ and μ but not the shape parameter ξ. It should be noted that in contrast to the block maxima some modifications are needed to point process-related techniques and we refer for the details to Longin 2016.

As was discussed, time dependence can be handled theoretically as well as practically in a rather satisfactory manner. This is unfortunately not generally the case for non-stationary series and a more pragmatic approach is usually adopted here instead. The basic idea is to use the standard extreme value models and change the underlying parameters. For example, variations through time in the observed process can be modeled as a linear trend in the location or scale parameter of the appropriate extreme value model, e.g.,

$$X_t \sim \mathrm{GEV}(\xi, \mu(t), \sigma(t)),$$

where

$$\mu(t) = \beta_0 + \beta_1 t$$

$$\sigma(t) = \beta_0 + \beta_1 t$$

for parameters β_0 and β_1 (these parameters can also be different for the location and scale parameter).

Generally speaking, there is usually insufficient evidence in data to support time-dependent models for the shape parameter ξ, though in principle this parameter can also be modeled. At least some sensitivity tests can be performed to see what eventual changes in the shape parameter would mean in regards to the probability of extreme events. This ends the section about extreme value theory and corresponding statistical techniques to estimate extreme events and we move on to risk management issues.

2.3 Risk Management Using Loss Distributions

In the previous sections, we introduced the term individual risk and defined it as a random variable represented through a distribution. We discussed that for the tails of the distribution, i.e., extremes, special techniques need to be applied to accurately estimate them. Having set up a distribution in this way, we defined a loss distribution to be a distribution with pure downside risk, e.g., losses. The loss distribution can be used to calculate various measures for risk management purposes. Most importantly in applied research is the expectation and some quantile measures such as the Value at Risk as they are very often used to set up risk management strategies as well, e.g., insurance schemes (see, for example, Heilmann 1988; Zweifel and Eisen 2012). However, there is a demand and supply side perspective to risk management that needs to be taken into account. In doing so one has to ask under which circumstances a decision-maker wants to get rid of his risks and under which circumstances someone would be willing to take his risk. This leads to decision theory and risk aversion which will be discussed next within the context of having a loss distribution available representing downside risk. The first goal in this section is to give an intuitive understanding of the risk premium, risk aversion, and utility concept and we follow the discussion related to Eeckhoudt and Gollier (1995). By using these concepts, we also take a look at the risk neutrality paradigm for governments and the violation of it in the case of extremes. We finally discuss the underlying assumption for risk spreading which naturally leads to dependency considerations that will be addressed in the next chapter.

To start with, consider a decision-maker which has some given initial wealth ω_0 and wants to take a look at his future wealth at the end of some time period $t > 0$, i.e., ω_t. In the easiest case, nothing happens in between and therefore the wealth at the beginning would be the same at the end. Hence, assume that something between the two time periods can happen. Either a gain or a loss could realize and in both cases, the initial wealth would be different compared to the wealth at the end. If the realization of a loss or a gain is random, which is often the case with investments or hazard event losses, it can, in principle, determined via a distribution function. In this case, the decision-maker could evaluate his future wealth through the use of some risk measures of the distribution (see Table 2.1). However, without any possibility to change either the distribution or applying risk management options to divert his risk he would need to stop at this point, i.e., he would know about the future risk and possible outcomes but could not do anything about it. Therefore, let us assume that there is also another distribution available of future risk which he could chose from (e.g., due to implementation of risk management options). The question we want to address is which options he should choose under which circumstances.

Let us therefore assume a decision-maker with initial wealth ω_0 and a random variable X that represents future loss and gains (for example, estimated using the techniques discussed before). As loss and gains are random also the future wealth ω_t will be random and $\omega_t = \omega_0 + X$. A classic example to start with is fire insurance. Assume for the moment that the decision-maker has a house worth 100.000 USD which with 1% probability could burn down (say till next year) and with 99% probability nothing happens. He wants to know if he should insure his house or not. For example, he calculates the expectation of his future wealth (e.g., next year) as $\mathbf{E}(\omega_f) = 100.000 * 0.99 + 0 * 0.01 = 99.000$ USD which shows that the risk to him, on average, is not that large. However, for indication what variability has to be expected around the mean, he also calculates the variance, which is $(100.000 - 99.000)^2 * 0.99 + (0 - 99.000)^2 * 0.01 = 99.000.000$ USD which is very large. This is a typical low probability high impact event scenario which characterizes extreme events. Let's assume he can buy fire insurance which would cost him 10.000 USD per year. There are now two possible future wealth situations, one where he bought fire insurance and one where he didn't. If he did not bought fire insurance the expectation would be still 99.000 USD and if he bought it he would have 90.000 USD. Just looking at the expectation of future wealth he would chose not to buy fire insurance. However, if he takes the variance into account he finds that in the non-insurance case he still has a variance of 90 million but for the insured case he does not have any risk (as the variance is zero). Focusing on the expectation for the evaluation of a risky situation is usually called using the *mean criterion*, in case that one also takes the variance into account and weight it, i.e., $\mathbb{E}(x) - k\sigma^2$ with $k = 0, 1, 2, \ldots$, this is usually called the *mean-variance criterion*. For our decision-maker for all cases with $k \geq 1$ buying insurance would be selected as most preferable option. In that regard, k can be seen as a kind of risk aversion parameter, i.e., the larger the k, the larger is the aversion to fluctuations around the mean (or risk).

The variance weights losses as well as gains equally. To account only for losses in a decision one can also use a tail measure in the mean-variance criterion, such a criterion is also sometimes called *safety-first criteria*. For example, let's assume another risky situation and a decision-maker who has an initial wealth of 100 USD is confronted to chose between two possible future scenarios A and B (Table 2.3)

Table 2.3 Two possible future scenarios and possible individual outcomes

Outcome A	Probability A	Outcome B	Probability B
−50	0.2	−150	0.01
4	0.3	10	0.49
10	0.5	30	0.5

Let's take a look at the criterion's introduced before. The expectation of scenario A is −3.8 while for scenario B it is 18.4. Hence, from an expectation criterion, one would chose scenario B. The variance for scenario A is around 540 and for scenario B it is 385. From an expectation-variance criterion again scenario B would be chosen if k = 1, as for scenario A it is −544.1 and for scenario B it is only −367. However, if one uses the semi-variance with the threshold set to zero (e.g., only losses which will bring the final wealth below 0) it would be zero for scenario A and 25 for scenario B. Hence, the mean-semi-variance criterion would lead to a selection of scenario A as it is larger (100) compared to scenario B (75).

The current discussion was focusing just on the distribution (and risk measures) of the random outcomes, however, already the famous St. Petersburg paradox indicated that it is not only about absolute losses or gains. The paradox uses a simple model of a fair coin toss, i.e., with 50% probability there will be tail and with 50% probability there will be face. Assume a game is played where a player would get 2 USD if in the first toss a tail would appear, 4 USD if a tail would appear only at the second toss, and so on, i.e., 2^n USD if tail just appears on the nth toss. The expectation for such a game is simply $\mathbf{E}(X) = 2 * \frac{1}{2} + 4 * \frac{1}{4} + ... = 1 + \cdots + 1 + \cdots = \infty$, hence one could assume that everybody would like to play this game as the expectation is ∞. However, in reality a player does not evaluate such a game simply through the expectation (or other criteria). Something is missing here to explain this behavior which was finally conceptualized as utility (see for a discussion Seidl 2013). Based on the St. Petersburg Paradox by Nicholas Bernoulli (1713) the concept of utility was formalized by Von Neumann and Morgenstern (1953). It can be used to model the behavior of decision-makers, e.g., risk aversion, in a very elegant way. Here, we assume for simplicity that the utility function of a decision-maker is continuous. Furthermore, we assume that the expected utility criterion will modify the values of ω_f to $U(\omega_f)$ but not transform the probabilities. Nevertheless, in many real-world cases also the probabilities could be, in principle, changed, e.g., for natural hazard events such as flooding by building dykes that can change the frequency and severity of events and therefore would affect the expected utility too. To avoid confusion in this section, we refrain including this possibility but will discuss this in some detail in the next two chapters. Let's call the evaluation criteria as **V** which is used to evaluate a risky situation expressed through the random variable X. Let's also assume we have a utility function available for a decision-maker which is strictly increasing and continuous (this is mainly assumed for simplicity reasons but can be extended to more complex utility functions rather easily). Based on the expectation formula to evaluate the expected utility of final wealth one has to evaluate

$$V(\omega_f) = \int_a^b U(\omega_0 + x)f(x)dx,$$

where a and b are the minimal and maximal values of the random variable X. In the next step, we want to answer the question on how much final wealth without the risky situation would lead to the same satisfaction as with an initial endowment and the risky situation. We call this final wealth the *certainty equivalent* ω^*. As we assumed that U is monotonic, it has an inverse and therefore

$$\omega^* = U^{-1}(U(\omega^*)) = U^{-1}(\int_a^b U(\omega_0 + x)f(x)dx).$$

In other words, a decision-maker is indifferent between a wealth of ω^* obtained with certainty and a risky situation composed of the initial wealth ω_0 and the random variable X. A difference in the utility function can be used to reflect a difference in the attitude toward risk. This fact can be used to serve as a measurement of the degree of risk aversion. For example, let's use as utility function of final wealth the square as well as the square root, i.e., $U(\omega_f) = \omega_f^2$ and $U(\omega_f) = \sqrt{\omega_f}$. Let's take as an example scenario A from Table 2.3. We first calculate for each possible future situation the final wealth and afterward take the square of it or the square root. Then we calculate the expectation for both utility functions. The expected utility for the first utility function is 9794 while for the second one it is 9.7. Taking the inverse this would lead to a certainty equivalent of 99 and 94, respectively. Hence, the decision-maker with the square utility function would be equally satisfied with having 99 USD instead of having 100 USD initial wealth and the risky situation, while the decision-maker with a square root utility would have a certainty equivalent of 94 USD. Note, the expectation without using a utility function is in between and around 96 USD. So somehow they evaluate the same situation differently, the first seems to evaluate a risky situation more positively and the later more negatively. In other words, one could be seen as risk seeker and one as risk avert. How to make this concept more tractable is most easiest when using the certainty equivalent as discussed next.

The certainty equivalent is the way toward defining risk attitudes and the risk premium plays a pivotal role here. First note that the decision-maker can consider what are the fair terms (to himself) of exchanging the risky situation (ω_0 and the random variable X) and certainty ω^*. The minimum price demanded by a decision-maker to sell his risky situation is called the *asking price* p_a defined as

$$p_a = \omega^* - \omega_0.$$

Every offer below this value would lead to rejecting the offer and therefore would not be sold by the decision-maker. Note that a negative asking price means that the decision-maker would be willing to pay anyone who is willing to take the risky situation off its hands. The concept of an asking price very much relates to the idea of insurance since he gets away with a risky situation. Conversely, there can be the case that a decision-maker (such as an insurance company) considers buying a risky situation. The maximum amount of money that a decision-maker would pay to get the risky situation is usually called the *bid price* p_b which is the solution to

$$U(\omega_0) = \left(\int_a^b U(\omega_0 + x - p_b) f(x) dx \right).$$

Hence, the value of p_b makes the decision-maker indifferent between his current situation and the situation of buying the risky situation. One could assume that p_a and p_b are the same for a given U, ω_0 and X, however, in reality this is not necessarily true (for example, the seller and buyer of the risky situation can have different utility functions).

Moving forward to the risk premium, we first want to note that for a decision-maker with a linear utility function in the final wealth the asking price is equal to the expected value of the risky situation $\mathbf{E}(X)$. Hence, a decision-maker with a linear utility function is using the mean value criterion for his evaluation. Sometimes it is said that the linearity of U implies *risk neutrality* as it cares only about the central aspects of the distribution without considering any other characteristics of risk. The case of risk neutrality is used as the foundation for defining risk aversion and risk loving. For risk neutrality, we say that the risk premium π is zero, where π is defined by

$$\pi = \mathbf{E}(X) - p_a.$$

We say that there is risk aversion if $\pi > 0$ as it implies that $p_a < \mathbf{E}(X)$ where the expected value of the risky situation would be the asking price *if* the individual were risk neutral. As π is greater zero this implies that the asking price of the risky situation is less than the asking price that would result from risk neutrality, hence the decision-maker does not value the risky situation. The same can be deducted for a risk lover. More formally, risk aversion and risk-loving behavior can be related to the shape of the utility function to be convex or concave.

Theorem 2.10 *A strictly increasing and strictly concave (convex) utility function of a decision-maker leads to a strict positive (negative) risk premium assigned to any risky situation.*

The proof is standard and based on Jensen's inequality. From the above theorem, various avenues can be taken to get a better insight into the risk premium and relative risk aversion and we refer to the many books already out there for getting a full picture (for example, the classic book by Eeckhoudt and Gollier 1995).

Especially when it comes to loss distributions one can assume for many decision-makers a risk-averse tendency, e.g., losses are assumed to cause negative utility. However, that may not always be the case if it is possible to spread risk sufficiently. In the case of the government, Arrow and Lind (1978) made an important contribution using the introduced concept of a positive risk premium for risk-averse decision-makers and due to its importance is discussed in more detail next.

Let's assume a government invests in a risky asset and expects a return Z which is a random variable with $\bar{Z} = \mathbf{E}(Z)$ and $X = Z - \bar{Z}$ with $\mathbf{E}(X) = 0$. Furthermore, assume identical individuals with given utility function U for their income. Additionally, assume that U is bounded, continuous, strictly increasing, and differentiable. Further assume a disposable income for all individuals to be i.i.d random variable B_0. If we consider a specific taxpayer and denote the fraction of his investment in the risky asset of the government by ε with $0 \leq \varepsilon \leq 1$ his disposable income given the government investment is equal to $B_0 + \varepsilon Z = B_0 + \varepsilon \bar{Z} + \varepsilon X$ which represents his share of Z. Using the expected utility criterion his future wealth is

$$\omega_f = \mathbf{E}(U(B_0 + \varepsilon \bar{Z} + \varepsilon X)).$$

We will show that the total of the risk premiums for all individuals' approaches zero and therefore the government can be assumed to be risk neutral as it needs to consider for its decision only the expectation of the net benefits $\bar{Z}$. In doing so, we first take the derivative of ω_f

$$\omega_f' = \mathbf{E}(U'(B_0 + \varepsilon \bar{Z} + \varepsilon X)(\bar{Z} + X))$$

and because the disposable income B_0 is independent of X it follows that also $U'(B_0)$ and X are independent and therefore $\mathbf{E}(U'(B_0)X) = \mathbf{E}(U'(B_0))\mathbf{E}(X) = 0$. Together with $\varepsilon = 0$ we get

$$\omega_f' = \mathbf{E}(U'(B_0)(\bar{Z} + X)) = \bar{Z}\mathbf{E}(U'(B_0)).$$

Note, from the definition of a derivative the formula above can be stated as

$$\lim_{\varepsilon \to 0} \frac{\mathbf{E}(U(B_0 + \varepsilon \bar{Z} + \varepsilon X) - U(B_0))}{\varepsilon} = \bar{Z}\mathbf{E}(U'(B_0))$$

and if we define $\varepsilon = 1/n$ (e.g., n is the number of tax payers in a country) one gets

$$\lim_{n \to \infty} n\mathbf{E}(U(B_0 + \frac{\bar{Z} + X}{n}) - U(B_0)) = \bar{Z}\mathbf{E}(U'(B_0)).$$

Let's assume that the individual decision-maker is risk averse, then we know that there exists a risk premium $\pi > 0$ so that it would be indifferent between paying the premium and accepting the risky situation with the investment, so

$$\mathbf{E}(U(B_0 + \frac{\bar{Z}+X}{n})) = \mathbf{E}(U(B_0 + \frac{\bar{Z}}{n} - \pi)).$$

As π is dependent on n (as it is related to the risk represented by $(1/n)X$) this shows that the $\lim_{n\to\infty} \pi = 0$, i.e., the costs of having the risky situation (through the investment B_0) are going to zero. However, it has to be shown that the total risk premium $n * \pi$ goes to zero as well. From the above we know that

$$\lim_{n\to\infty} n\mathbf{E}(U(B_0 + \frac{\bar{Z}}{n} - \pi) - U(B_0)) = \bar{Z}\mathbf{E}(U'(B_0)$$

and that $\frac{\bar{Z}}{n} - \pi$ goes to zero as n tends to infinity. Hence,

$$\lim_{n\to\infty} \frac{\mathbf{E}(U(B_0 + \frac{\bar{Z}}{n} - \pi) - U(B_0))}{\frac{\bar{Z}}{n} - \pi} = \mathbf{E}(U'(B_0)) > 0$$

which finally leads to (by dividing the two equations above)

$$\lim_{n\to\infty} (\bar{Z} - n\pi) = \bar{Z}$$

and therefore $\lim_{n\to\infty} n\pi = 0$. Hence, according to Arrow-Lind a government can make risk neutral investments, or in other words, only need to choose risky situations based on the mean evaluation criterion. However, in Hochrainer and Pflug (2009) it was shown that in the case of extreme events the total premium may not go down to zero and therefore governments should behave risk averse in these cases. Their argument is based on the violation of the assumption of having a continuous utility curve and that in cases of disasters utility may show a kink which essentially makes it not possible anymore to derive at a zero risk premium. The rather technical analysis is skipped here, and we come back to the special case of governments as decision-makers in a later chapter.

Up to now, we assumed that for the evaluation of the asking price or the premium the whole distribution is looked at. This does not necessarily need to be the case and very often a decision-maker just wants to get rid of part of the risky situation. In modern societies a risk-averse decision-maker typically manage his individual risks through insurance or, more generally, through diversification. The so-called Excess-of-Loss (XL) insurance contract represents a broad case of possible insurance schemes, including deductibles as well as a cap on possible claim payments. In other words, XL insurance contracts are a combination of contracts with a deductible, and contracts with an upper bound. A typical excess-of-loss contract requires the insured person to retain a specific level of risk with the insurer covering all losses between an attachment point a and an exit point d, this function is illustrated in Fig. 2.10.

Analysis of how premiums and optimal insurance contracts can be determined is an active research field (see the related references). We just want to note here

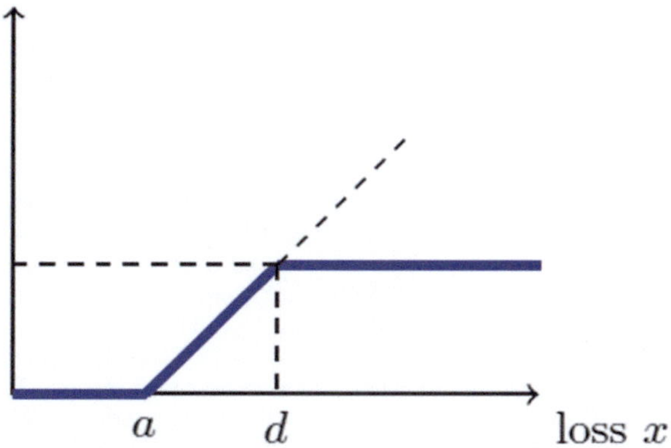

Fig. 2.10 Example of an typical XL insurance contract

that while moderate individual risks, such as car accidents, can be handled efficiently in this way, diversification gets increasingly difficult for extreme risks (Kessler 2014; Linnerooth-Bayer and Hochrainer-Stigler 2015). In such cases, risk diversification has to be done through reinsurance or catastrophe bonds, and details for such instruments can be found in Hochrainer (2006) and Cardenas et al. (2007). All of these instruments, while quite complicated, are based on the ideas presented above. Another risk management option commonly applied for frequent and extreme risks is through risk reduction usually analyzed through cost-benefit analysis, also here we refer for the interested reader to the manifold literature out there (Mechler 2016). We will come back to the management of individual risk in more detail in the next chapter that presents a risk-layer approach and we apply these concepts in the application chapter as well. The basic ideas were introduced in this chapter and we move forward to the dependencies between risks as the essential ingredient for systemic risk analysis discussed in the next chapter.

References

Arrow KJ, Lind RC (1978) Uncertainty and the evaluation of public investment decisions. In: Uncertainty in economics. Elsevier, pp 403–421

Bauer H (2011) Measure and integration theory, vol 26. Walter de Gruyter

Cardenas V, Hochrainer S, Mechler R, Pflug G, Linnerooth-Bayer J (2007) Sovereign financial disaster risk management: the case of Mexico. Environ Hazards 7(1):40–53

Castillo E, Hadi AS, Balakrishnan N, Sarabia J-M (2005) Extreme value and related models with applications in engineering and science. Wiley

Coles S, Bawa J, Trenner L, Dorazio P (2001) An introduction to statistical modeling of extreme values, vol 208. Springer

Eeckhoudt L, Gollier C (1995) Risk: evaluation, management and sharing. Harvester Wheatsheaf

Embrechts P, Klüppelberg C, Mikosch T (2013) Modelling extremal events: for insurance and finance, vol 33. Springer Science & Business Media

Fisher RA, Tippett LHC (1928) Limiting forms of the frequency distribution of the largest or smallest member of a sample. In: Mathematical proceedings of the cambridge philosophical society, vol 24. Cambridge University Press, pp 180–190
Gnedenko B (1943) Sur la distribution limite du terme maximum d'une serie aleatoire. Ann Math, 423–453
Gumbel EJ (1958) Statistics of extremes, p 201. Columbia University Press, New York
Heilmann WR (1988) Fundamentals of risk theory. Verlag Versicherungswirtsch
Hochrainer S (2006) Macroeconomic risk management against natural disasters. German University Press
Hochrainer S, Pflug G (2009) Natural disaster risk bearing ability of governments: consequences of kinked utility. J Nat Disaster Sci 31(1):11–21
Hochrainer S, Mechler R, Pflug G, Lotsch A (2008) Investigating the Impact of climate change on the robustness of index-based microinsurance in Malawi. The World Bank
Hogg RV, Klugman SA (2009) Loss distributions, vol 249. Wiley
Jenkinson AF (1955) The frequency distribution of the annual maximum (or minimum) values of meteorological elements. Q J Royal Meteorol Soc 81(348):158–171
Katz RW, Parlange MB, Naveau P (2002) Statistics of extremes in hydrology. Adv Water Resour 25(8–12):1287–1304
Kessler D (2014) Why (re) insurance is not systemic. J Risk Insur 81(3):477–488
Klugman SA, Panjer HH, Willmot GE (2012) Loss models: from data to decisions, vol 715. Wiley
Kotz S, Nadarajah S (2000) Extreme value distributions: theory and applications. World Scientific
Leadbetter M R, Lindgren G, and Rootzén H (1983) Extremes and Related Properties of Random Sequences and Processes. Springer, New York
Linnerooth-Bayer J, Hochrainer-Stigler S (2015) Financial instruments for disaster risk management and climate change adaptation. Climatic Change 133(1):85–100
Longin F (ed) (2016). Extreme events in finance: A handbook of extreme value theory and its applications. John Wiley & Sons
Malevergne Y, Sornette D (2006) Extreme financial risks: from dependence to risk management. Springer Science & Business Media
Mechler R (2016) Reviewing estimates of the economic efficiency of disaster risk management: opportunities and limitations of using risk-based cost-benefit analysis. Natural Hazards 81(3):2121–2147
Mechler R, Bouwer LM, Linnerooth-Bayer J, Hochrainer-Stigler S, Aerts JC, Surminski S, Williges K (2014) Managing unnatural disaster risk from climate extremes. Nature Climate Change 4(4):235
Nicholson S, Klotter D, Chavula G (2014) A detailed rainfall climatology for Malawi. Southern Africa Int J Climatol 34(2):315–325
Panjer HH (2006) Operational risk: modeling analytics, vol 620. Wiley
Pflug G, Roemisch W (2007) Modeling, measuring and managing risk. World Scientific
Pickands J III et al (1975) Statistical inference using extreme order statistics. Ann Stati 3(1):119–131
Reiss R-D, Thomas M, Reiss R (2007) Statistical analysis of extreme values, vol 2. Springer
Roussas GG (1997) A course in mathematical statistics. Elsevier
Sachs L (2012) Applied statistics: a handbook of techniques. Springer Science & Business Media
Seidl C (2013) The St. Petersburg paradox at 300. J Risk Uncertain 46(3):247–264
Von Neumann J, Morgenstern O (1953) Theory of games and economic behavior. Princeton University Press
Wilks DS (1990) Maximum likelihood estimation for the gamma distribution using data containing zeros. J Climate 3(12):1495–1501
Zweifel P, Eisen R (2012) Insurance economics. Springer Science & Business Media

Chapter 3
Systemic Risk and Dependencies

In Chap. 2, we discussed individual risk defined as a random variable characterized by its distribution function. We further discussed how distributions can be estimated based on past observations. We treated extreme risk estimation separately within this discussion as a theory of its own was needed for such kind of risks, namely, Extreme Value Theory (EVT). We discussed how distributions can be constructed where one part of the distribution represents non-extremes, estimated via classic estimation techniques, while for the other part, namely, the tail of the distribution, extreme value statistics is applied to accurately estimate extreme events. In this chapter, we now assume that such a distribution is already in place and in case of pure downside risk, e.g., losses, we call the corresponding distribution a loss distribution.

This chapter now focuses at so-called systemic risks. In what follows, we repeat part of the discussion already done in the introduction chapter to set up the stage. Before the term systemic risk can be introduced in a satisfactory manner, it is first necessary to establish an understanding of what is meant by a system in our context, what kind of systems exist (or may be assumed to exist) in reality, and how such systems can be analyzed. We try to keep this discussion rather short as the main focus of this chapter is on the dependency between loss distributions which simplifies the answer to these questions. A prominent social science perspective on systems is given by Luhmann (1995) (see for an introduction also Luhmann et al. 2013; Baecker 2016) and in regard to complexity we refer to Helbing (2007) and more generally to Thurner et al. (2018). The focal point for our analysis of a system here is based on a network perspective. This will naturally lead to the idea to interpret a so-called copula as a network property.

To start very simplistic, we first want to note that if a system consists of only one individual element that is "at risk" it simply can be treated as individual risk as done in Chap. 2 before. Even in the case of many unconnected individuals within the system, one can treat each element as carrying individual risks only. The situation changes if some kind of interaction between the elements of a network has to be assumed (Fig. 1.4). If an element is (somehow) dependent or influenced by another element, an additional dimension comes into the picture which was not addressed yet, namely,

© Springer Nature Singapore Pte Ltd. 2020

S. Hochrainer-Stigler, *Extreme and Systemic Risk Analysis*, Integrated Disaster Risk Management, https://doi.org/10.1007/978-981-15-2689-3_3

an interaction or dependency dimension. These interactions can produce higher order effects, usually subsumed under the term emergence (Barabási and Albert 1999), and surprisingly such higher order systems can behave quite differently than the elements they consist of (see again Luhmann 1995). As we can observe in reality everything can be (at least conceptually, it does not necessarily need to be the case that something like a "system" actually exists, see a discussion about this issue in Luhmann et al. 2013) seen as either an element within a system or a system itself consisting of elements (see for an ontological discussion Elias 1987). As Fig. 1.3 indicates, various different stages of integration can be looked at from this perspective.

It is obvious that the complexity of the systems in Fig. 1.3 increases from the bottom to the top. One possibility to define complexity is in terms of the "structuredness" of a system: properties of the composite elements are increasingly more dependent on the organization of the elements and less dependent on the actual properties of the isolated elements (see Thurner et al. 2018). Consequently, from a systems perspective at some point, the focus on isolated observed characteristics is less adequate than the focus on the organization between elements. Methodological wise, it implies that the modeling of interdependent relations and processes becomes increasingly necessary. Ideally, such analysis could be performed either theoretical or empirically, depending on the question, scale, data availability, and nature of the problem (e.g., systemic risk) at hand. However, in reality, due to lack of information and understanding large uncertainties must be (sometimes) expected in regard to the likelihood of events as well as about potential outcomes. We refer to the literature for an introduction to these issues and details (see Florin et al. 2018) and in the context of our approach to Sect. 3.3.

Modeling wise and again generally speaking, there are two distinct approaches for studying systems and systemic risk within a network approach such as depicted in Fig. 1.4. Either one starts with a network and some characterization of it *or* one defines the relevant elements and interaction behaviors and let the network build up by themselves. In the latter case, this includes so-called agent-based modeling approaches but can also be done in other ways, for example, using differential equations as a starting point. At some point for both approaches, one ends up with a network which usually has some stable regions as well as ones with high systemic risk. As already introduced in Chap. 1, systemic risk is also called network risk (Helbing 2013). The realization of systemic risk leads, by definition, to a breakdown or at least major dysfunction of the whole system (Kovacevic and Pflug 2015). Systemic risk can realize exogenously, e.g., through outside attacks, or endogenously, e.g., due to sudden failures in network nodes. Essentially, systemic risk is usually due to complex cascading effects among interconnected agents ultimately leading to collective losses, dysfunctions, or collapses. A lot of current research of systemic risk is within the financial domain, especially banking systems, however, already in the past and now increasingly again also other research domains focus on systemic risk (see Hochrainer-Stigler et al. 2019). Indeed, systemic risk (similar to extreme risk) has a wide area of application and will get even more important in our increasingly complex (and therefore interdependent) world (Centeno et al. 2015). The diverse field of applications of systemic risk also indicates that many different phenomena (e.g., too big to fail, too interconnected to fail, keynote species) on how systemic risk can realize are possible.

Taking a network perspective on systems, as it is widely accepted in the systemic risk literature, in what follows a system will be defined as a set of interconnected individual elements which are "at risk". As in this book a distribution approach is adopted, meaning that individual risks are treated as random variables with a corresponding distribution function, the interconnections between them need to be modeled through appropriate statistical dependence measures which is the central topic of this chapter and which leads to our suggested approach for integration of extreme and systemic risk by reinterpreting Sklar's theorem. To start with, we will introduce and discuss the most important dependence measures in the next section.

3.1 Dependence Measures

For simplicity, we first assume that we have only two random variables X and Y that are following a given cumulative distribution function, notated as F_X and F_Y. We want to know to which extent both "move together". We first want to define the covariance between the two random variables

$$\begin{aligned}\mathbf{Cov}(X,Y) &= \mathbf{E}[(X-\mathbf{E}(X))(Y-\mathbf{E}(Y))]\\ &= \mathbf{E}[XY - X\mathbf{E}(Y) - \mathbf{E}(X)Y + \mathbf{E}(X)\mathbf{E}(Y)]\\ &= \mathbf{E}(XY) - \mathbf{E}(X)\mathbf{E}(Y) - \mathbf{E}(X)\mathbf{E}(Y) + \mathbf{E}(X)\mathbf{E}(Y)\\ &= \mathbf{E}(XY) - \mathbf{E}(X)\mathbf{E}(Y).\end{aligned}$$

In terms of the corresponding distributions, the formula above can be stated also as (sometimes referred to as Hoeffding's lemma)

$$\mathbf{Cov}(X,Y) = \int_{-\infty}^{\infty}\int_{-\infty}^{\infty}(P(X \le x, Y \le y) - F_X(x)F_Y(y))dxdy.$$

An unbiased estimator of the covariance is the so-called sample covariance calculated as $\frac{1}{n-1}\sum_{i=1}^{n}(x_i - \bar{x})(y_i - \bar{y})$, where $\bar{x}$ and $\bar{y}$ are the sample mean and n is the number of observations. As already discussed in Chap. 2, the covariance is instrumental for providing test statistics and confidence intervals. Note, the value of the covariance is dependent on the scale used. However, if one normalizes the covariance appropriately, it can be used as a measure of dependency between two random variables that can capture the degree of *linear* relationship, both strength and direction wise. This is the so-called Pearson correlation coefficient ρ defined as

$$\rho = \frac{\mathbf{Cov}(X, Y)}{\sqrt{\sigma^2(X)\sigma^2(Y)}}.$$

With σ^2 being the variance of the corresponding random variables, the values of ρ are always between $[-1, 1]$ indicating strong negative or strong positive dependence. The Pearson correlation coefficient is one of the most often used dependence measures in statistics. As an classical example, Fig. 3.1 shows the dependency between the Canadian inflation (based on CPI) and short-term interest rates from 1954 to 1994.

The graphs on the upper left and lower right corners in Fig. 3.1 are showing the histograms for both variables, while the two other figures are showing the corresponding scatterplot of past observations. The number indicated on the upper left-hand side is the Pearson correlation coefficient calculated to be 0.74, which could be interpreted as a strong positive correlation between the two random variables.

While very useful in practical applications, it should be kept in mind that it is a linear dependence measure (see the straight line in the scatterplot in Fig. 3.1), and

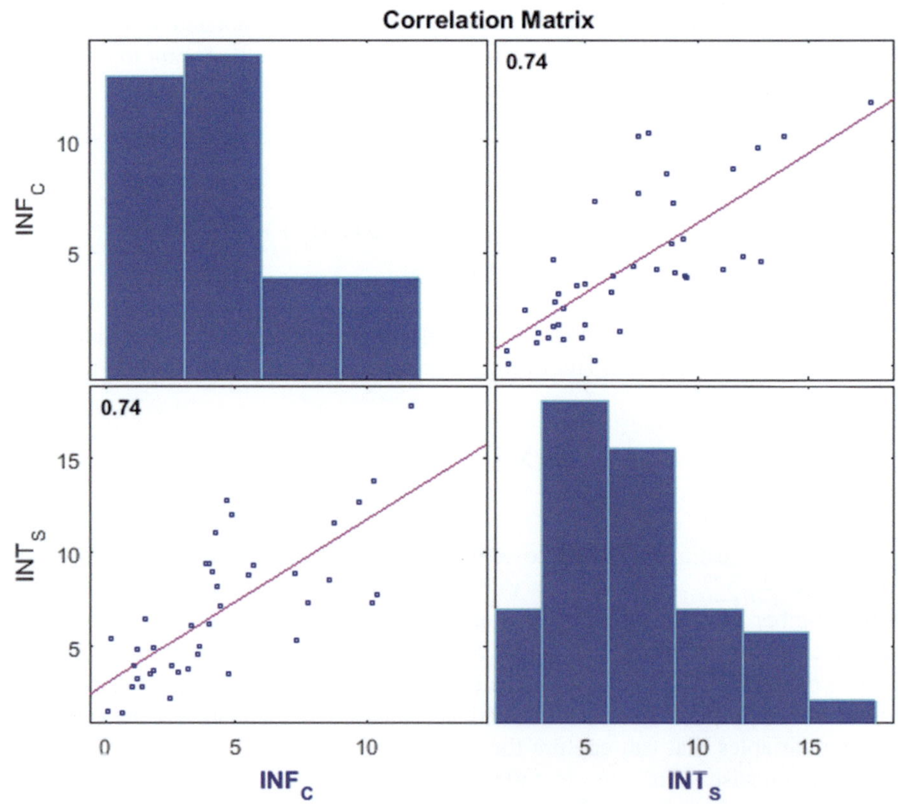

Fig. 3.1 Scatterplot and Pearson correlation of Canadian inflation and interest rate

hence does not take into account any eventual differences in the correlation strength in the upper or lower parts of the distributions of the random variables. Additionally, one has to be careful when working with fat-tailed data as the variances of X and Y have to be finite for the Pearson correlation coefficient to be defined. Another limitation is that ρ is not invariant under strictly increasing transformations. Finally, uncorrelation (i.e., $\rho = 0$) does not mean independence, while the converse holds true.

Due to these limitations, there is also another very prominent dependence measure available which is used very often in applied research, the so-called Kendall's rank correlation coefficient τ. It also plays an important role for copulas. Kendall's tau is a non-parametric measure and as the name indicates it is a rank correlation measure. Usually, standard empirical estimators of rank correlation are calculated by looking at the ranks of the data, i.e., the focus is on the ordering of the sample and not the actual numerical values (see McNeil et al. 2015; Denuit et al. 2006). The underlying concept is concordance. For example, two points (x_1, y_1) and (x_2, y_2) are said to be concordant if $(x_1 - x_2)(y_1 - y_2) > 0$ and discordant if $(x_1 - x_2)(y_1 - y_2) < 0$. Based on this concept, if we assume that (X, Y) and (X', Y') (defined as an independent copy with the same distribution but independent of the first), Kendall's rank correlation is simply the probability of concordance minus the probability of discordance for these pairs:

$$\tau(X, Y) = P((X - X')(Y - Y') > 0) - P((X - X')(Y - Y') < 0). \quad (3.1)$$

Hence, if Y tends to increase with X, we also expect the probability of concordance to be high relative to the probability of discordance while if Y tends to decrease with increasing X we expect the opposite. It is sometimes convenient to write Kendall's tau in terms of expectation which is

$$\tau(X, Y) = \mathbb{E}(\text{sign}((X - X')(Y - Y'))). \quad (3.2)$$

As a simple example to show the usefulness of Kendall's tau, let's assume that we have paired observations from two random variables from an unknown distribution (Table 3.1) and want to calculate the Pearson correlation coefficient as well as Kendall's tau.

For an unbiased estimator of the Pearson correlation $\hat{\rho}$, one also needs the sample mean for both observations, i.e., $\bar{x} = \frac{1}{n}\sum_{i=1}^{n} x_i$ and $\bar{y} = \frac{1}{n}\sum_{i=1}^{n} y_i$. Then

$$\begin{aligned}
\hat{\rho} &= \frac{\frac{1}{n-1}\sum_{i=1}^{n}(x_i - \bar{x})(y_i - \bar{y})}{\sqrt{\frac{1}{n-1}\sum_{i=1}^{n}(x_i - \bar{x})^2}\sqrt{\frac{1}{n-1}\sum_{i=1}^{n}(y_i - \bar{y})^2}} \\
&= \frac{-43.5}{2.16 * 22.93} \\
&= -0.88.
\end{aligned}$$

Table 3.1 Observations used for Pearson and Kendall correlations

x	1	2	3	4	5	6	7
y	64	32	16	8	4	2	1

Note, while between x and y a functional relation could be actually established, as the Pearson correlation coefficient is a linear measure, it gives a correlation coefficient below 1. This is not the case with Kendall's tau which can be calculated as

$$\tau = \frac{2K}{n(n-1)}$$

with K being the total number of concordance or discordances and n being the number of total observations. For example, for x_1, x_2, y_1, y_2 in Table 3.1, the number K can be calculated as $(1–2)(64–32) = -1 * 32 = -32$ which is smaller than zero and therefore one would get $K = -1$ for this comparison, doing this for all possible permutations and summing them up will finally lead to $K = -21$ and therefore $2 * (-21)/(7 * 6) = -1$, i.e., full negative correlation.

The last dependence measure we introduce here before the copula is presented is the so-called "tail dependence coefficient" (see, for example, Nelsen 2007) which aims at summarizing the dependence in the joint tail of the multivariate distribution. The term tail dependence usually refers to the dependence between the random variables in the tails. In the case of bivariate distributions, it measures the dependence in the upper right or lower left quadrant of the unit cube. The coefficient is defined as the limit of the conditional probabilities of quantile exceedances.

Definition 3.1 Let X and Y be continuous random variables with distribution F_X and F_Y, respectively. Provided that the limits exist, the coefficients of lower (λ_L) and upper tail dependence (λ_U) of X and Y are defined as

$$\lambda_U = \lim_{t \to 1^-} P[Y > F_Y^{-1}(t) | X > F_X^{-1}(t)]$$
$$\lambda_L = \lim_{t \to 0^+} P[Y \leq F_Y^{-1}(t) | X \leq F_X^{-1}(t)].$$

If $\lambda_U(X, Y) \in (0, 1]$, then it is said that the random variables are upper tail dependent. If $\lambda_L(X, Y) \in (0, 1]$, then it is said that they are lower tail dependent. If $\lambda_U(X, Y) = 0$, they are said to be upper tail independent (the same for $\lambda_L(X, Y) = 0$ which is said that they are lower tail independent).

The upper and lower tail-dependent coefficients will be helpful to describe copulas and also Kendall's tau can be closely linked to copulas as well. To introduce copulas (see also a good example in Hofert et al. 2018 for an introduction into copulas), we first look at two extreme cases of dependency between two random variables, i.e., complete independence and complete dependence. Let's assume we have two random variables X_1 and X_2 representing losses and we want to know the total losses, i.e.,

$X = X_1 + X_2$. If both distributions are assumed to be independent than (statistical) convolution can be applied to produce the joint distribution and the density of the total loss is equal to

$$f(x) = f_1 \star f_2 = \int f_1(x-z) f_2(z) dz, \tag{3.3}$$

where f_1 and f_2 are the marginal densities of X_1 and X_2 and $\star$ is the convolution operator (Sachs, 2012). In the case they are fully dependent, the two distributions can be simply summed up to derive at the joint distribution (Hochrainer-Stigler et al. 2014a). A copula can provide a detailed model between these two extreme cases (see Timonina et al. 2015). Statistically, a copula is a particular type of multivariate distribution function. A multivariate distribution function, say G of a d-dimensional random vector $X = (X_1, \ldots, X_d)$, can be defined as

$$G(x) = P(X \leq x) = P(X_1 \leq x_1, \ldots, X_n \leq x_d), \quad x = (x_1, \ldots, x_d) \in R^d.$$

As each F_j of X_j can be recovered by G through $F_j(x) = G(\infty, \ldots, x_j, \ldots, \infty)$, $x_j \in R$, the distributions $F_1, \ldots, F_d$ are usually called univariate margins of G or the marginal distributions of X. Using this notation, we are able to give the classic definition of a copula.

Definition 3.2 A d-dimensional copula C is a distribution function on $[0, 1]^d$ with standard uniform marginal distributions.

The requirement that the margins are standard uniform is actually arbitrary, however, still a sensible decision (see Hofert et al. 2018 for a short discussion on this issue). In the case of a multivariate copula, the copula will be denoted as $C(\mathbf{u}) = C(u_1, \ldots, u_d)$. Simply put, Definition 3.2 says that a copula maps the unit hypercube (somehow) into the unit interval. While we will only look at established copulas throughout the text, it should be noted for completeness reasons that the following three properties must hold (see McNeil et al. 2015):

(i) $C(u_1, \ldots, u_d)$ is increasing in each component u_i, $i = 1, \ldots, d$.
(ii) $C(1, \ldots, u_i, 1, \ldots, 1) = u_i$, for all $i \in \{1, \ldots, d\}$, $u_i \in [0, 1]$
(iii) for all $(a_1, \ldots, a_d), (b_1, \ldots, b_d) \in [0, 1]^d$ with $a_i \leq b_i$ one has

$$\sum_{i_1=1}^{2} \cdots \sum_{i_d=1}^{2} (-1)^{i_1+\ldots+i_d} C(u_{1i_1}, \ldots, u_{di_d}) \geq 0,$$

where $u_{j1} = a_j$ and $u_{j2} = b_j$ for all $j \in \{1, \ldots, d\}$.

The first property was already discussed as a property in the context of distribution functions, while the second property is simply the statement about uniform marginals. The third property makes sure that the probability is always non-negative. We also want to quickly note the Frechet-Hoeffding bounds due to its importance in copula theory. Define W and M as

$$W(u) = \max\left\{\sum_{j=1}^{d} u_j - d + 1, 0\right\}$$

$$M(u) = \min_{1 \le j \le d}(u_j)$$

with $u \in [0, 1]^d$, then for any d-dimensional copula C the following inequalities hold.

Theorem 3.1 *For every copula, the following bounds hold:*

$$W(u) \le C(u) \le M(u) \tag{3.4}$$

In proceeding, one can get easily lost in the details of copulas which we want to avoid here and rather focus on application aspects and most important attributes of copulas. Hence, we omit completely the discussion of "generator functions" which are usually introduced in the context of copulas (see Trivedi et al. 2007) and we also do not discuss exotic and very special cases of copulas such as quasi-copulas and semi-copulas (see Durante and Sempi 2015). There are also many different parametric copula families available, including elliptical copulas such as the Gaussian copula, Student's copula, or Archimedean copulas (see Joe 2014). The Archimedean copulas and extreme value copula families are the most important ones in our case and will be looked at in more detail. For all others, we refer to Klugman et al. (2012) and Sachs, 2012. Among the large number of copulas in the Archimedean copula family, four of them are especially useful and will be used throughout the book and therefore discussed in more detail next.

To show the nature of copulas, we will heavily rely on graphs. The graph of a two-dimensional copula is a continuous surface within the unit cube I^3 which lies between the graphs of the Frechet-Hoeffding bounds. A very useful graph for a given copula is also through a so-called contour diagram which shows level sets, i.e., sets in the unit cube given by $C(u, v) =$ constant for selected constants. For example, Fig. 3.2 shows the upper and lower limit cases of a copula with the help of a contour diagram. However, scatterplots of simulated observations from copulas as well as corresponding narratives serve equally well for the illustration of different copula types and will be the method of choice in this and the next section.

We start with an introduction of some important Archimedean copulas. The discussion and examples given follow Trivedi et al. (2007). For instance, various studies about the joint survival of husband/wife pairs have found some non-linear behavior with strong tail dependence. This so-called "broken heart" syndrome in which individual's death substantially increases another individuals' probability to die is the prime candidate to be modeled through copulas such as the so-called Clayton copula (see Trivedi et al. 2007). *The Clayton copula* is an asymmetric Archimedean copula and exhibits strong left tail dependence and relatively weak right tail dependence.

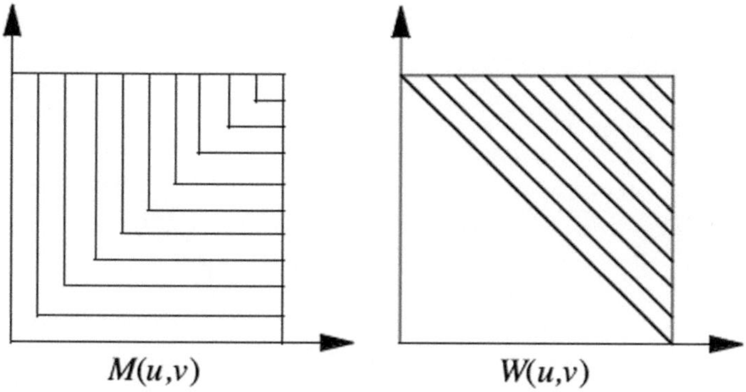

Fig. 3.2 Examples of upper and lower limits of copulas

Informally, tail dependence occurs when the dependency between random variables increases as one gets into the tails of the distributions. For example, the Clayton copula exhibits left tail dependence which can be illustrated easily if one simulates (discussed further down below) observations from this copula and produces a scatterplot afterward (see Fig. 3.3). The smaller the values on both axes, the less the spread of observations between them, or in other words the stronger the dependency between them. Contrary, the larger the values on both axes, the more spread in the observations can be observed and one speaks of weak right tail dependence. The Clayton copula can be stated as

$$C_\theta^C(u, v) = \left(u^{-\theta} + v^{-\theta} - 1\right)^{-\frac{1}{\theta}}, \qquad 0 < \theta < \infty.$$

As indicated in Fig. 3.3, it can account only for positive dependence; however, as θ approaches zero it can also be used to model independence. It can also be noted that as θ approaches ∞ it attains the Frechet-Hoeffding upper bound (but never the lower bound). Various dependency strengths can be modeled via the Clayton copula parameter, and some selected scatterplots (as indicated based on random sampling procedures discussed further down below) for visualization of these strengths in regard to the parameter choice can be found in Fig. 3.3 too.

A symmetric Archimedean copula is the so-called *Frank copula* ($\theta \in [-\infty, \infty]$, with $\theta \to 0$ corresponds to independence) which is of the following type:

$$C_\theta^F(u, v) = -\frac{1}{\theta} \ln\left(1 + \frac{(e^{-\theta u} - 1)(e^{-\theta v} - 1)}{e^{-\theta} - 1}\right). \tag{3.5}$$

It has been widely used in empirical applications in the past; however, it only exhibit weak tail dependence (with the strongest dependence centered in the middle). Important is also that the dependence is symmetric in both tails, and both Frechet-Hoeffding

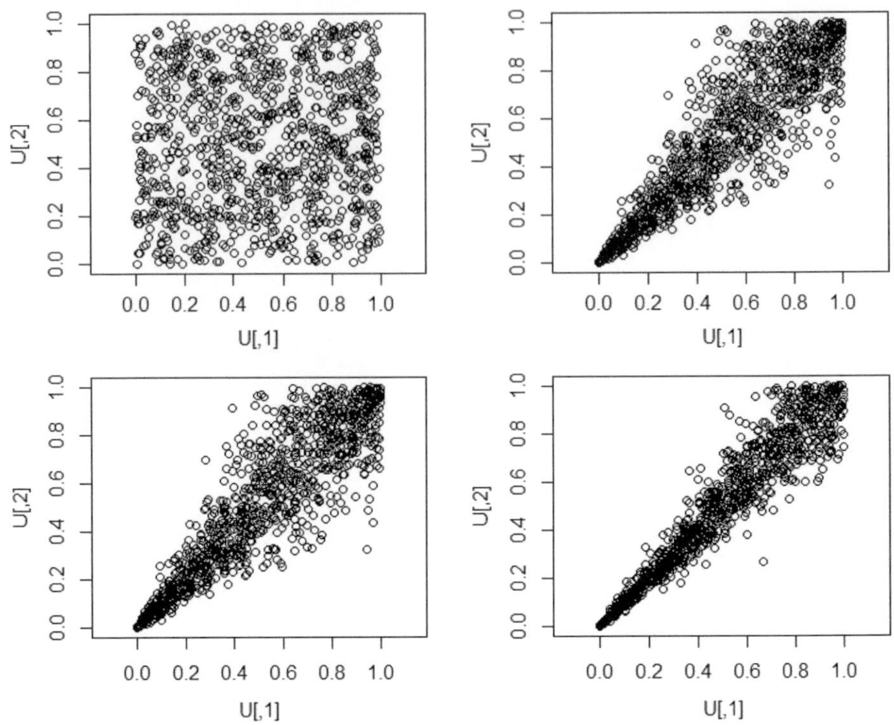

Fig. 3.3 Example of Clayton copula strengths, theta=[0,4,4,8]

bounds are within the permissible range of dependence. For illustration purposes, Fig. 3.4 shows the Frank copula for two different parameter settings.

Another important copula is the *Gumbel copula* which is again an asymmetric Archimedean copula, exhibiting greater dependence in the right tail and relatively weak left tail dependence. Hence, such a copula can be used as a candidate model if outcomes are strongly correlated at high values but less correlated at low values.

$$C_\theta^G(u,v) = \exp\left[-\left[(-\ln u)^\theta + (-\ln v)^\theta\right]^{\frac{1}{\theta}}\right] \tag{3.6}$$

with ($\theta \geq 1$). Figure 3.5 shows a Gumbel copula with dependence parameter $\theta = 5$. Note, values of θ of 1 or ∞ correspond to independence and the Frechet upper bound.

Finally, we introduce the *Joe copula*. It is similar to the Gumbel copula but allows even stronger upper tail dependence. It is the independence copula given $\theta = 1$ and converges to the comonotonicity copula as $\theta \to \infty$ (Ruppert and Matteson 2015). The Joe copula is defined as

$$C_\theta^J(u,v) = 1 - [(1-u)^\theta + (1-v)^\theta - (1-u)^\theta(1-v)^\theta]^{\frac{1}{\theta}} \tag{3.7}$$

for $\theta \in [1, \infty]$.

Fig. 3.4 Example of two Frank copulas, theta=[4,15]

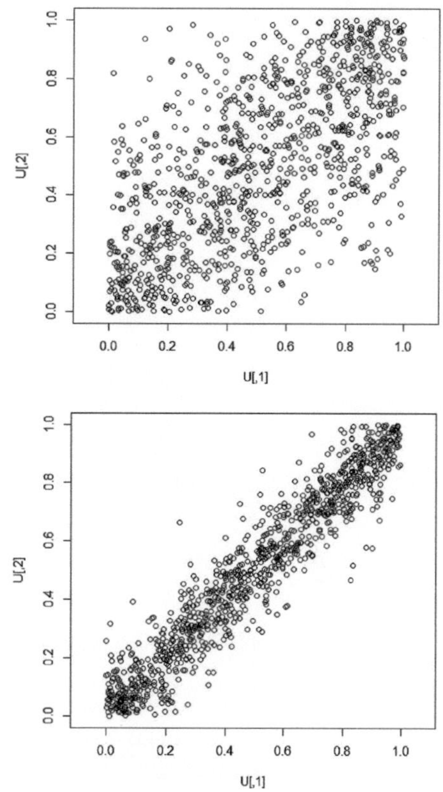

Due to the strong upper tail dependence of the Joe copula, it is closer to the Clayton copula in the left tail-dependent case. However, it is also possible to flip the Clayton copula so that the new copula exhibits greater dependence in the positive tail. The flipped Clayton copula was, for example, discussed in Timonina et al. (2015) and can be written as

$$C_\theta^{FC}(u, v) = u + v - 1 + \left[(1-u)^{-\theta} + (1-v)^{-\theta} - 1\right]^{-\frac{1}{\theta}}.$$

As we will see, not only the copulas themselves are important but also the conditional copulas. Conditional copulas can be determined as the partial derivative of $C_\theta^{(\cdot)}(u, v)$ over v, i.e.,

$$C_\theta^{(\cdot)}(u|v = v_0) = \lim_{\Delta u \to 0} \frac{C(u + \Delta u, v) - C(u, v)}{\Delta u} = \frac{\partial C_\theta^{(\cdot)}(u, v)}{\partial v}|_{v=v_0}.$$

Conditional copulas will be instrumental for simulation purposes. For example, using the inverse of the conditional copula with respect to u, one can generate random

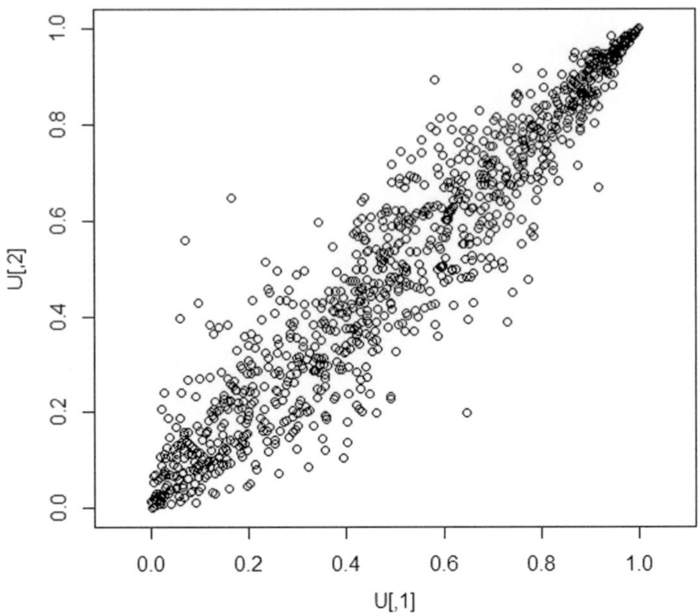

Fig. 3.5 Example of a Gumbel copula, theta=5

numbers from the conditional distribution by sampling from the uniform distribution for v. Furthermore, they are very useful for the application of multivariate copula approaches as we will see further down below. The Clayton as well as the flipped Clayton and Frank copulas are directly invertible (for other copulas, it can be rather complicated), and hence the generation of corresponding samples is not very complicated for these copulas. For example, the conditional transform for the copula generation of the Clayton copula (see for example Trivedi et al. 2007) is

$$u = \left[v^{-\theta} \left(r^{-\frac{\theta}{1+\theta}} - 1 \right) + 1 \right]^{-\frac{1}{\theta}}. \tag{3.8}$$

For the conditional flipped Clayton copula, it is (see Timonina et al. 2015)

$$u = 1 - \left[1 + (1 - v)^{-\theta} (r^{-\frac{\theta}{1+\theta}} - 1) \right]^{-\frac{1}{\theta}} \tag{3.9}$$

and for the conditional Frank copula the conditional transform is

$$u = -\frac{1}{\theta} \ln \left(1 + \frac{r(1 - e^{-\theta})}{r(e^{-\theta v} - 1) - e^{-\theta v}} \right), \tag{3.10}$$

with r as a uniform random variable on the interval $(0, 1)$. For the case of the Gumbel copula, the conditional copula is not directly invertible. However, as discussed in Timonina et al. (2015), one can consider the partial derivative of the Gumbel copula over v:

$$\frac{1}{C_\theta^G(u,v)}\frac{\partial C_\theta^G(u,v)}{\partial v}=\frac{(-\ln v)^{\theta-1}}{v}\left[(-\ln u)^\theta+(-\ln v)^\theta\right]^{\frac{1-\theta}{\theta}}$$

and making the following change of variables

$$w=\exp\left[-\left[(-\ln u)^\theta+(-\ln v)^\theta\right]^{\frac{1}{\theta}}\right],$$

so that the differential is

$$\frac{\partial C_\theta^G(u,v)}{\partial v}=\frac{(-\ln v)^{\theta-1}}{v}w(-\ln w)^{1-\theta}, \tag{3.11}$$

which is a concave function in case that w is between 0 and 1 and $\theta \geq 1$ and therefore can be solved numerically. The corresponding algorithm can be found in Timonina et al. (2015), and a more detailed discussion can be found in Trivedi et al. (2007).

The consequences of simulating from a conditional copula are illustrated in the two graphs in Fig. 3.6. There 1000 samples of v having set u to 0.05 as well as u to 0.95 are shown. As can be seen, the values of v are not randomly distributed but follow a pattern according to the underlying copula type, in this case, the Clayton copula.

This ends our discussion about conditional copulas, and we move forward on how the copula parameters can actually be estimated. Indeed, as in the individual and extreme risk case in Chap. 2, there is the need to estimate the unknown parameters and different approaches such as the standard parametric maximum-likelihood estimator, method-of-moment estimator, semiparametric pseudo-maximum-likelihood estimator, or the minimum-distance estimator were developed accordingly. We refer to Trivedi et al. (2007) for an overall introduction into these techniques and to Joe (2014) for a theoretical analysis. A very prominent technique is to estimate the copula parameters with the so-called Inference Functions for Margins (IFM) method (Joe 1997) as it gives consistent estimates of the dependence parameter (under some regularity conditions). The basic idea is that the parameters of the marginal distributions (see Sect. 3.3) are estimated in a first step and one determines the copula parameters in the second step (Joe 2014). In other words, given N observations, one estimates in a first step with ML methods (see Chap. 2) the parameters $\hat{\beta}_i, i = 1, \ldots, T$ of T univariate marginal distributions (e.g., individual risks). In the second step, the copula parameters are estimated given the previous estimates: If c_θ is the density of a copula C_θ, the ML estimate of the parameter $\hat{\theta}$ is

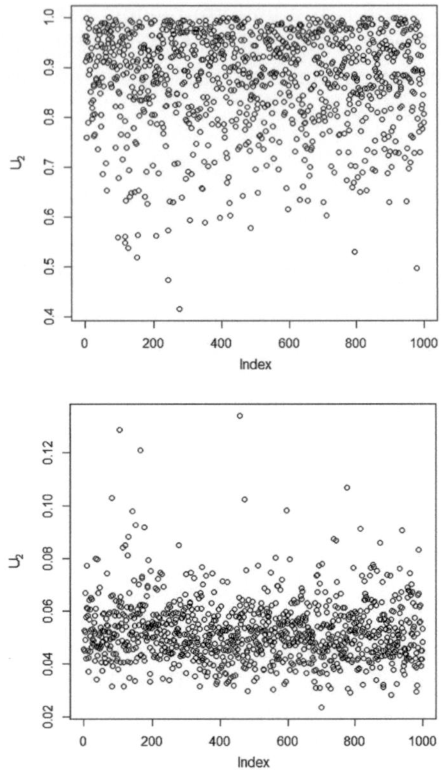

Fig. 3.6 Example of conditional Clayton copula simulations based on u to be 0.05 and 0.95

$$\hat{\theta} = \text{argsup}_{\theta \in \Theta} \sum_{i=1}^{N} \log c_{\theta}(F_1(x_{1,i}), \ldots, F_T(x_{T,i})),$$

where $F_1(x_1, i)$ is the marginal distribution with fitted parameter $\hat{\beta}_1$ and observation i, etc. One major disadvantage of this approach is the loss of efficiency in estimation as in the first step the dependence between the marginal distributions is not considered. However, it is also possible to estimate the parameter of the copula without assuming some specific parametric distribution functions of the marginals and to use instead pseudo-observations. Sometimes, this approach is referred to as the Canonical Maximum-Likelihood (CML) method (also pseudo-log-likelihood method). In case of the CML method, the data needs first to be transformed into uniform variates $(\hat{u}_1^i, \ldots, \hat{u}_T^i)$ (so-called pseudo-observations), and

$$\hat{\theta} = \text{argsup}_{\theta \in \Theta} \sum_{i=1}^{N} \log c_{\beta}(\hat{u}_1^i, \ldots, \hat{u}_T^i).$$

Table 3.2 Relationship between copula parameter and Kendall's tau for selected Archimedean copulas

Clayton	$\tau = \frac{\theta}{\theta+2}$
Frank	$\tau = 1 + \frac{4}{\theta}\left(\frac{1}{\theta}\int_0^{\theta}\frac{t}{\exp(t)-1}dt - 1\right)$
Gumbel	$\tau = \frac{\theta-1}{\theta}$

The range of applicability of the CML method is far greater compared to other methods such as the method of moments, as long as the density c_θ can be evaluated (and therefore making it possible to maximize the logarithmic pseudo-likelihood). As starting values for such optimization, pair-wise sample versions of Kendall's τ, for example, can be used. For the bivariate copula case, the rank correlation coefficient τ can be related to the copula parameters (see, for example, Table 3.2). For a more detailed discussion on these relationships, we refer to Durante and Sempi (2015).

As was indicated at the beginning of this section besides the Archimedean copula family, there is also the family of extreme value copulas which we want to look at now (also in light of the discussion in Sect. 2.2). They are derived from the dependence structure of the multivariate GEV distribution as discussed in the previous chapter. In more detail, consider N i.i.d. n-dimensional random variables $\mathbf{X}_k = (X_{k,1}, \ldots, X_{k,n})$ for $k = 1, \ldots, N$ with distribution function F. Define the component-wise maxima as

$$M_{j,N} = \max_{1 \leq k \leq N} X_{k,j}. \tag{3.12}$$

If for a suitable chosen norming sequence $(a_{k,N}, b_{k,N})$ the limit distribution exists, then it is given by

$$C(H_{\xi_1}(x), \ldots, H_{\xi_n}(x)), \tag{3.13}$$

where H_ξ is the already defined GEV distribution and C is an extreme value copula. Hence, this class of copulas is defined in terms of the scaling property of extreme value distributions, i.e., a copula is an extreme value copula if it satisfies

$$C(u_1^\alpha, \ldots, u_n^\alpha) = (C(u_1, \ldots, u_n))^\alpha, \qquad u \in [0, 1]^n, \alpha > 0. \tag{3.14}$$

Due to the scaling property, it can be shown that an extreme value copula is max-stable. The Gumbel copula discussed before as an example of the Archimedean copula is also an extreme value copula. However, there are no other Archimedean extreme value copulas. Another extreme value copula is the Galambos copula which is of the form (see Klugman et al. 2012)

$$C(u, v) = uv \exp(((-lnu)^{-\theta} + (-lnv)^{-\theta})^{-1/\theta}). \tag{3.15}$$

There exists also an asymmetric version of the Galambos copula, which has the form

$$C(u, v) = uv \exp(((-\alpha lnu)^{-\theta} + (-\beta lnv)^{-\theta})^{-1/\theta}), \tag{3.16}$$

where the one-parameter version is obtained by setting $\alpha = \beta = 1$. Also, here many different copulas exist, including the Tawn copula or the BB5 copula. There are also combinations of the Archimedean and extreme value copulas into single class of copulas possible, called Archimax copulas which includes the BB4 copula, see for more information, Klugman et al. (2012) and Nelsen (2007).

As indicated by the discussion, many different copula functions exist and there is the question how to select the most appropriate one, e.g., which copula fits the data best. During the exploratory phase, one may graphically examine dependence patterns, e.g., through scatterplots. For example, if the X and Y are highly correlated in the right tail than a flipped Clayton copula or a Joe copula might be an appropriate copula. Most formal goodness of fit tests are implemented mainly for one-parameter Archimedean copula families and multivariate elliptical copulas. The basic idea here is to apply some global distance measures between the model-based and empirical distribution but this approach has some serious limitations. We refer for a comprehensive discussion to Joe (2014). Generally speaking, there are not many adequate tests available and one is rather using the Akaike Information Criterion (AIC) or the Bayesian Information Criterion (BIC). These are information-based criteria based on the likelihood function with different penalty functions for overfitting. Usually, both are used to assess the model fit and to compare different kinds of models. Given k is the number of estimated parameters in a model, $ln(x)$ is the maximized log-likelihood value, and N is the number of observations, the AIC and BIC are defined as

$$\begin{aligned} AIC &= -2ln(x) + 2k \\ BIC &= -2ln(x) + kln(N). \end{aligned}$$

As a general rule and given a set of candidate models, the models with the smaller AIC or BIC value can be considered to be a better fit compared to the others with larger values. There are also penalized likelihood measures and corresponding test available too (see Trivedi et al. 2007).

It must be noted that we only gave a rough overview of possible copula types, inference as well as estimation techniques and we referred for a comprehensive treatment to the books already mentioned above. The goal here was merely to give an intuitive understanding and some basic information about measuring and modeling copulas including a presentation of some very important copula types used in applications. While the current discussion focused mostly on the bivariate case, we extend now our discussion in the next section and focus on techniques for multivariate dependency modeling through copulas.

3.2 Multivariate Dependence Modeling with Copulas

Important applications of copulas require to go beyond the bivariate copula modeling case and to establish multivariate copula models instead. However, the construction of multivariate copulas is difficult. Few of the bivariate cases can be easily extended to the multivariate case. For example, the Gumbel copula (see a more detailed discussion in Nelsen 2007) has the following extension to a $(2^n - n - 1)$ parameter family of n-copulas with $n \geq 3$:

$$C(u) = u_1 u_2 \ldots u_n \left[1 + \sum_{k=2}^{n} \sum_{1 \leq j_1 \ldots\ldots\ldots j_k \leq n} \theta_{j_1 j_2 \ldots j_k} (1 - u_{j_1}) \ldots (1 - u_{j_k}) \right].$$

while for multidimensional copulas closed form solutions are available as above, they have many parameters to be estimated *or* they are restricted as they use only one (or two) parameter for modeling the joint multidimensional dependency structure. For example, Timonina et al. (2015) discussed the following multivariate flipped Clayton copula which depends only on one parameter θ:

$$C_\theta(u_1, \ldots, u_n) = \sum_{i=1}^{n} u_i - 1 + \left[\sum_{i=1}^{n} (1 - u_i)^{-\theta} - 1 \right]^{-1/\theta}.$$

Hence, several methods were developed to overcome this inflexibility of multivariate copulas, but these techniques are usually omitted within an introduction into copula modeling as the theory gets very fast very complicated. Here, we will focus instead on very practical approaches which are especially useful for our network perspective to analyze systemic risk including hierarchical and vine-copula approaches (the discussion follows Gaupp et al. 2017).

A quite reason but already widely used approach are *vine* copulas also referred to as pair copula constructions. The name comes from the fact that they combine a cascade of conditional and unconditional bivariate copulas of arbitrary types into a n-dimensional copula. The pair-copulas are ordered according to tree structures with the three most common ways of ordering including D-vines (e.g., line trees), C-vines (e.g., star trees), and the most flexible R-vines. Figure 3.7 shows an example of a 5-dimensional C-vine and a R-vine tree.

Figure 3.7 indicates that if each copula pair can be chosen independently the number of pair-wise copulas is already very large and therefore also the possible dependence structures in C- or R-vines (even in the case of small dimensions). For example, a d-dimensional copula can have $\frac{d(d-1)}{2}$ bivariate copulas that can be estimated in $\frac{d!}{2}$ possible C-vine trees and $\frac{d!}{2} 2^{\binom{d-2}{2}}$ possible R-vine trees (see Czado et al. 2013). More generally, a vine structure has $(d - 1)$ trees with nodes N_i and edges E_{i-1} joining the nodes.

The actual tree structures are built by considering the proximity condition which states that if an edge connects two nodes in tree $j+1$, the corresponding edges in tree j share a node. Formally, a regular vine tree sequence on n elements is defined (see Czado et al. 2013) as a set of linked trees $V = (T_1, \ldots, T_{n-1})$ with

1. T_1 is a tree with set of nodes denoted by $N_1 = \{1, \ldots, n\}$ and set of edges by E_1
2. For $i = 2, \ldots, n-1$, T_i is a tree with nodes $N_i = E_{i-1}$ and edge set E_i
3. The proximity condition: For $i = 2, \ldots., n-1$, if $a = \{a_1, a_2\}$ and $b = \{b_1, b_2\}$ are two nodes in N_i connected by an edge, then exactly one of the a_i equals one of the b_i

The R-vine copula is then defined by (F, V, B) where F is a vector of distribution functions, V is an n-dimensional R-vine tree, and B is a set of bivariate copulas. As already indicated, a special case of regular vines is the C-vine or canonical vines where each tree $T_i, i = 1, \ldots, n-1$ has an unique node (called root node) that is connected to $n-1$ edges, or in other words is connected to all other nodes of the tree. For D-vines also called drawable vines, all nodes in a tree are connected at most to two edges.

There are many different selection approaches available on how to select actually the structure of a tree. For example, in a study by Aas et al. (2009), the focus was on the strongest bivariate dependencies (measured through Kendall's tau) to build up the tree structure. However, the possible edge weighting can be manifold, including, for example, the p-value of a selected goodness of fit test or by simply taking the copula parameter θ as a measure of strength. Very often a maximum spanning tree approach is adopted using Kendall's τ as edge weights for R-vines (see, for example, Dissmann et al. 2013). In this case, for each tree, the spanning tree which maximizes the sum of absolute Kendall's τ's is selected which boils down in solving the following optimization problem:

$$\max \sum_{e_{ij} \in T} |\hat{\tau}_{ij}|,$$

with e_{ij} denoting the edges and T is the spanning tree. The optimization starts from the first $T_1 = N_1, E_1$ and is solved for each next tree where the nodes withthe strongest

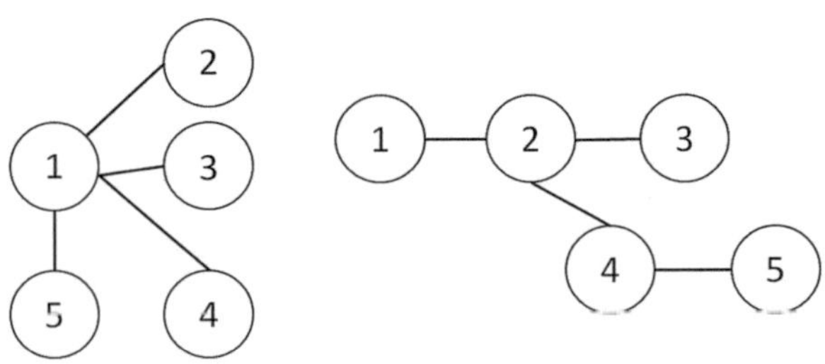

Fig. 3.7 C- and R vine tree example

dependency with the other nodes are identified. After finding the tree structure, the best fit bivariate copula families can be selected for each pair of variables $e_{j,k}$ in the optimal spanning tree. In that way, the spanning tree and selection of copulas can be separated. As will be discussed later, such an approach can also be used to determine the network structure of a system at hand.

Due to the still large number of parameters needed for estimating the optimal spanning tree and corresponding copulas, other approaches such as *hierarchical copulas* can be used instead. One major benefit is the decrease of number of parameters to be estimated, e.g., for a d-dimensional copula only $d-1$ parameters need to be estimated. As the name already indicates, the structure is based on a specific hierarchy between pairs. This can be done by several methods, including minimax approaches (e.g., selecting the structure based on the strongest dependencies between pair-wise copulas within a given set of nodes, see Timonina et al. 2015) or based on some additional information (e.g., riverine structures, see Lugeri et al. 2010). Figure 3.8 shows an example of a hierarchical structure which is using nested pair-wise copulas, e.g.,

$$C_{12,23,36,45,56} = C_{35}(C_{12,23}, C_{45}). \tag{3.17}$$

Summarizing the techniques on how to build a multivariate copula model using spanning trees, four steps have to be taken (according to Gaupp et al. 2017): The first step involves the selection of an adequate copula structure. The selection can be based on expert knowledge (e.g., river branches in the case of flood risk) or can be found by applying minimax, C-vine or R-vine approaches. In step two, appropriate copulas have to be selected which can be done through visualization using scatterplots in the bivariate case or analytically through different goodness of fit tests such as the Kendall (Genest and Rivest 1993) or the Vuong and Clarke tests (Clarke 2007). Afterward (step three), the copula parameters are estimated and in step four, the models can be evaluated with criterions such as the already introduced AIC or BIC criterion.

Given a structure of dependency has been established and the copulas and parameters are selected from a family of copulas usually one wants afterward to simulate random variables from these copulas, for example, for stress-testing investigations or sensitivity tests (see also the applications in Chap. 4). One especially useful way for the simulation of copulas within spanning tree approaches is by using the conditional copulas. The method is based on the observation that

$$\begin{aligned} P(U_1 \le u_1, \ldots, U_n \le u_n) = {} & P(U_n \le u_n | U_1 = u_1, \ldots U_{n-1} = u_{n-1}) \\ & * P(U_1 \le u_1, \ldots U_{n-1} \le u_{n-1}), \end{aligned}$$

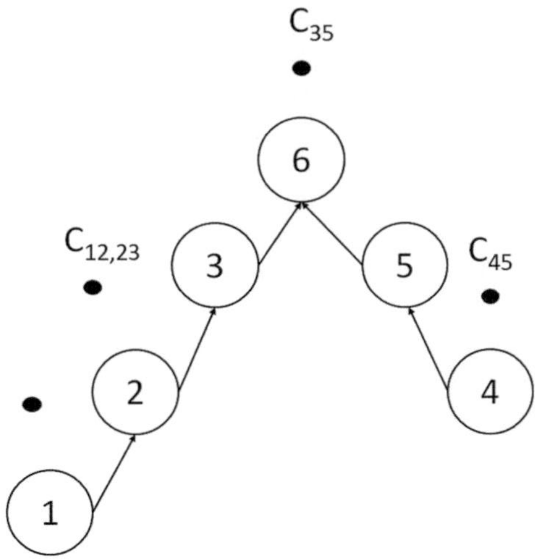

Fig. 3.8 Hierarchical copula example

which leads recursively to

$$
\begin{aligned}
P(U_1 \le u_1, \dots, U_n \le u_n) = {} & P(U_n \le u_n | U_1 = u_1, \dots U_{n-1} = u_{n-1}) \\
& * P(U_{n-1} \le u_{n-1} | U_1 = u_1, \dots U_{n-2} = u_{n-2}) \\
& * P(U_{n-2} \le u_{n-2} | U_1 = u_1, \dots U_{n-3} = u_{n-3}) \\
& \dots \\
& * P(U_2 \le u_2 | U_1 = u_1) * P(U_1 \le= u_1).
\end{aligned}
$$

As a copula is a distribution function, one can use this observation to get

$$C(u_1, \dots, u_n) = C_n(u_n | u_1, \dots, u_{n-1}) \dots C_2(u_2 | u_1) * C_1(u_1),$$

with

$$C_k(u_k | u_1, \dots, u_{k-1}) = \frac{\partial_{u_1} \dots \partial_{u_{k-1}} C_k(u_1, \dots, u_k)}{\partial_{u_1} \dots \partial_{u_{k-1}} C_{k-1}(u_1, \dots, u_{k-1})},$$

where the conditional copulas can be evaluated as described above. Hence, to simulate n random variables with a given copula C, one just has to

1. generate n uniform and independent random variables $v_1, \dots, v_n$
2. set $u_1 = v_1$
3. set $u_2 = C_2^{-1}(v_2 | u_1)$

   $\dots$
4. set $u_n = C_n^{-1}(v_n | u_1, \dots, u_{n-1})$.

This algorithm is especially useful for Archimedean copulas, and we give two examples of such a simulation approach for the Frank and Clayton copula (see Trivedi et al. 2007). To simulate Frank's copula, one can use the following algorithm:

1. generate two uniform and independent random variables v_1, v_2
2. set $u_1 = v_1$
3. set $u_2 = -\frac{1}{\theta} ln \left(1 + \frac{v_2(1-e^{-\theta})}{v_2(e^{-\theta v_1}-1)-e^{-\theta v_1}}\right)$

For the simulation of a Clayton copula with positive parameter θ, the following algorithm can be used:

1. generate two uniform and independent random variables v_1, v_2
2. set $u_1 = v_1$
3. set $u_2 = (v_1^{-\theta}(v_2^{-\theta/(\theta+1)} - 1) + 1)^{-1/\theta}$

This ends our short discussion on the topic of measuring and modeling dependencies and we want to proceed to discuss how a joint framework for individual and systemic risk analysis can be established. The focal point is now on Sklar's theorem, and we embed its discussion within a system's context as discussed in the introduction section.

3.3 A Joint Framework Using Sklar's Theorem: Copulas as a Network Property

Up to now two ingredients for our suggested integration of extreme and systemic risk analysis have been, quite independently, established. In Chap. 2, we discussed how to measure, model, and manage individual risk and focused on distributions. In Sects. 3.1 and 3.2, we discussed how to measure and model interactions between random variables through dependence measures, focusing on copula approaches. As already explained at the beginning of this chapter, we will take a network (Fig. 1.4) approach for defining a system, i.e., a system is defined as a set of interconnected elements. These elements can be "at risk", and we use a (loss) distribution to represent individual risk. In other words, we treat the elements within a system as random variables. Furthermore, in this chapter, we have introduced and focused on copulas which are special cases of multivariate distribution functions and we showed how to measure and model them. Based on distributions on the one hand and copulas on the other hand, the core idea of how to integrate both is through Sklar's theorem which is discussed next. When it comes to the discussion of copula approaches, the theorem is usually introduced at the very beginning; however, here we postponed it to emphasize now both concepts of dependence and individual risk within our network perspective. In a nutshell, Sklar's theorem (Sklar 1959) provides a one-to-one relationship between multivariate distributions and copulas. We suggest that Sklar's theorem can be reinterpreted from a network perspective, treating copulas as a network property and elements within the network which are at risk as random variables. In that way both, individual and systemic risks, can be analyzed independently as well as jointly.

Without exaggeration, it can be acknowledged that Sklar's theorem is one of the main reasons why copula approaches became famous and Sklar's theorem is now well known in the literature. As suggested in Fig. 1.4 from a system or network perspective, one can also interpret the marginal distributions as nodes and the copula as a model for determining the dependency between the nodes within the network. As will be discussed, this perspective enables a joint assessment of individual and systemic risk within a common framework over different scales. It is worthwhile to note that in doing so nothing new is established in terms of the theorem itself but rather a new perspective on the theorem is taken forward. To the author's best knowledge, the suggested perspective of a copula as a network property and its usefulness for the integration of individual and systemic risk analysis was given first in Hochrainer-Stigler (2016) and the comprehensive presentation in this chapter follows the arguments made in Hochrainer et al. 2018e. Due to its central importance, we start our discussion with Sklar's theorem first. As in the other sections also here we assume absolute continuous distributions for all cases where random variables are considered.

Theorem 3.2 *Let $F_X \in R^2$ be a 2-dimensional distribution function with continuous marginals $F_{X_1}(x_1)$ and $F_{X_2}(x_2)$. Then, there exists a unique copula C such that for all* x $\in R^2$

$$F_X(x_1, x_2) = C(F_{X_1}(x_1), F_{X_2}(x_2)). \tag{3.18}$$

Conversely, if C is a copula and $F_{X_1}(x_1)$ and $F_{X_2}(x_2)$ are distribution functions, then the function F_X defined by (3.18) is a bivariate distribution with margins $F_{X_1}(x_1)$ and $F_{X_2}(x_2)$.

Proof Let C be the joint distribution for the couple $F_{X_1}(x_1)$ and $F_{X_2}(x_2)$, that is,

$$\begin{aligned} C(u_1, u_2) &= P(F_{X_1}(X_1) \leq u_1, F_{X_2}(X_2) \leq u_2) \\ &= P(X_1 \leq F_{X1}^{-1}(u_1), X_2 \leq F_{X2}^{-1}(u_2)) \\ &= F_X(F_{X1}^{-1}(u_1), F_{X2}^{-1}(u_2)), \end{aligned}$$

then Sklar holds as

$$\begin{aligned} F_X(x_1, x_2) &= P(X_1 \leq x_1, X_2 \leq x_2) \\ &= P(F_{X1}(X_1) \leq F_{X1}(x_1), F_{X2}(X_2) \leq F_{X2}(x_2)) \\ &= C(F_{X1}(x_1), F_{X2}(x_2)). \end{aligned}$$

Hence, the dependence structure between random variables is entirely and uniquely described by the copula C and separated from the individual risk measured through marginal distributions. We assumed above that the distribution functions are continuous but it should be mentioned that for the discrete case there is more than one way possible for writing the dependency structure. While the copula presentation

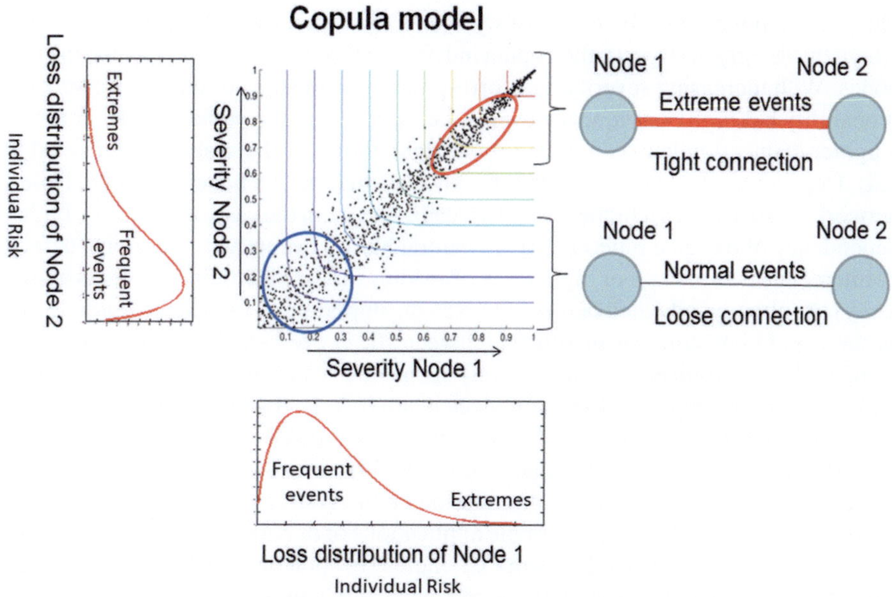

Fig. 3.9 Illustrating copulas and marginal distributions for normal and extreme events. *Source* Adapted based on Hochrainer-Stigler (2016) and Hochrainer-Stigler et al. (2018e)

would still hold in this case, C is no longer unique and usually is referred to as a "possible" copula (see for different proofs of Sklar's theorem Durante and Sempi 2015). It should also be noted that the manner how both random variables might behave together is captured by the copula regardless of the scale in which each of the random variables might be measured. This is another feature which is very beneficial for systemic risk purposes as some systemic risk may occur due to only relatively small losses in a specific node. For illustration purposes of the main idea and following partly the discussion in Hochrainer-Stigler (2016) and Hochrainer-Stigler et al. (2018e), Fig. 3.9 conceptually shows the reinterpretation of the copula concept as stated above interpreting the two random variables as two single nodes (i.e., the individual risks in terms of marginal distributions) and the copula as interaction model between the two nodes. For better illustration of the tails (e.g., extremes), a loss density distribution for the individual nodes is selected.

Let's take a closer look at this idea by starting with the right-hand side of Fig. 3.9. There are different strengths of connections assumed through the copula, for example, a loose connection between node 1 and node 2 for normal events (in our case frequent events) and a tight connection during extreme events (i.e., events in the tail of the distribution, see also the discussion in Chap. 2). As indicated, in the middle of Fig. 3.9 through samples of observations from a given copula (in this case the flipped Clayton), lower levels of u_1 and u_2 are dependent in a weaker sense than higher levels. During "normal" events (lower left corner of the scatterplot and marked in blue), corresponding to the low severity events for nodes 1 and 2 (or in other words where the density is largest), the copula indicates a loose connection between the two nodes. With increasing severity and corresponding to the tails of the two nodes (low probability but high consequence events, i.e., extremes), the copula model enforces a tighter connection between them (upper right corner of the scatterplot, marked in red). Depending on the research domain, such a situation can be reinterpreted from a network perspective as a higher risk of spreading (e.g., in the case of diseases), higher dependency of extreme outcomes (e.g., natural hazard events across regions, default of interconnected banks), or higher risk of loadings (e.g., electricity networks within a city). As already indicated before but given the importance, we want to reiterate that as the copula has uniform marginals between [0, 1] even small disturbances (e.g., compared to losses in another node level) in one node can result in a tight connection with another node which subsequently can result in large disturbances on the system level. Importantly, using the simulation approaches as discussed in Sect. 3.2 and having the copula as well as individual risk in the form of loss distributions available, one can also estimate now a loss distribution on the system level as well.

To show the usefulness and possible applications of our approach more clearly, we present next three superficial network examples (for practical real-world examples, we refer to the detailed applications presented in Chap. 4). The reader may have noticed that up to now we have defined individual risk as well as system-level risk through loss distributions and copulas. We have not discussed yet at what level of losses on the individual level as well as on the system level one could say that extreme risk or systemic risk will realize. As already indicated in the introduction section, this very much depends on how the given elements in the system as well as the system level itself can cope with eventual losses. The concept of coping capacity is very closely linked to the more broader term resilience and needs, due to its complexity, a discussion of its own (done in the next section). For simplicity, we first define systemic risk as in the case of individual risk through risk measures, e.g., Value at Risk (example 3 below), or through a threshold assumption (example 2 below). In more detail, in the first example, we will simply focus on the calculation of individual- and system-level risk. The second example takes a risk management approach to individual and systemic risk. Finally, the third example discusses a more complex multi-node network to show the influence of individual failure to systemic risk realization.

To start with, let's assume first a system which consists of two interconnected elements, i.e., we view the system as a two-node network. Furthermore, let's assume that the individual risk of each element in the network can be modeled through a Gamma distribution with mean 10 and variance of 20. This is just an arbitrary assumption and reflects the situation of individual risk such as in Fig. 3.9. Additionally, let's assume that tail dependence is according to a Joe copula. We already introduced contour plots for the Frechet-Hoeffding bounds above and show now such plots in Fig. 3.10 for the Joe copula. As one can see, quite different dependency strengths can be modeled through this copula, with a low dependence shown on the left-hand figure and a high dependence on the right-hand side.

Hence, a Joe copula can enforce a tighter connection in the tail which can increase losses during extreme events considerably. For example, let's look at the probability that losses in both nodes will be below 10, or in other words system-level losses would be 20 in that case. In the case of independence, the joint probability that losses for both individual risks are below this level is simply $F_X(X_1 \leq 10, X_2 \leq 10) = F_{X1}(10) * F_{X2}(10) = 0.56 * 0.56 = 0.31$, while assuming the Joe copula with parameter 9 one would get 0.52. Hence, the probability of system-level losses of 20 (or below) in that setting is severely underestimated for the independent case. In other words, such joint individual loss events will happen with much larger probability in the dependent system compared to the independent system. While in the independent case such a situation only happens with around 30% probability in the dependent system, it will happen with around 50% probability. Through simulation, one can determine all possible individual loss situations explicitly incorporating their dependency (such as exemplified here) and one is therefore able to determine a loss distribution on the system level as well, as discussed next.

The second example should demonstrate how one can use the suggested approach for risk management purposes for individual and systemic risks (we refer here to the example already introduced in the introduction section in Chap. 1). Let's assume a decision-maker which is responsible for a system that consists of two elements at risk. Let's further assume loss data is available for both elements and that losses in element (or node) 1 can be modeled with a Gamma distribution with shape parameter 1 and scale parameter 2 and losses in element (or node) 2 can be modeled with a Gamma distribution with shape parameter 7.5 and scale parameter 1. Note, while both risks are using the same distributional model, losses in the tails are quite different. For example, losses with a 100-year return period are corresponding to 9.2 and 15.3 for node 1 and node 2, respectively. Hence, extreme losses are much higher for node 2 compared to node 1, and also the shape of the distribution is quite different (see the distributions in Fig. 1.7).

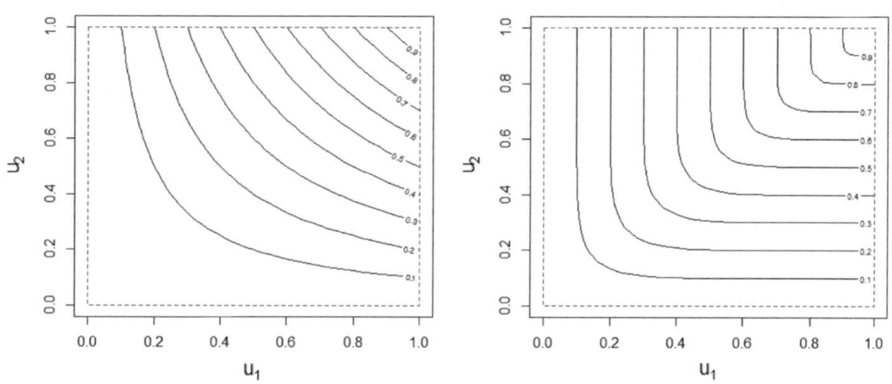

Fig. 3.10 Contour plots for Joe copulas with dependency parameters set to 1.5 and 9

The loss data from both nodes can be looked at together with different plots, e.g., through a scatterplot as depicted in Fig. 1.7. A decision-maker on the system level would notice that there seems to be a non-linear increase in both losses the larger the individual losses are. However, both nodes have quite different shapes and magnitudes, and therefore it is not immediately clear for the decision-maker on how an actual relationship between the two distributions can be established. Let's assume therefore that the decision-maker applies the copula techniques discussed before and finds out that a flipped Clayton copula with parameter 5 would be an appropriate model. Figure 1.8 shows the corresponding copula with univariate margins in the form of the already introduced scatterplots. The copula shows upper tail dependence, while no strong dependencies in the region of frequent events can be observed.

Having a copula and distributions for the system established, the decision-maker can apply simulation techniques to calculate a loss distribution on the system level. For example, he can use the estimated flipped Clayton copula parameter to simulate (as discussed in the previous section) 100.000 copula-based observations for both nodes, e.g., $(u_i, v_i), i = 1, \ldots, 100.000$. Using the inverse of the Gamma distribution (see Sect. 3.4), the losses for both nodes can be determined as well, e.g., $(F_1^{-1}(u_i), F_2^{-1}(v_i))$. In that way, the decision-maker has now coupled pairs of 100.000 simulated losses for both nodes exhibiting tail dependence. If the decision-maker is summing up the losses in both nodes for each sample (e.g., $F_1^{-1}(u_i) + F_2^{-1}(v_i)$), he now has 100.000 losses on the system level too. He could now also estimate a loss distribution on the system level using this sample. Assume that by using this data the decision-maker finds out (e.g., looking at the empirical distribution) that a 100-year event on the system level would cause losses of about 22 say, currency units. Let's assume for the moment that he has a budget of 20 currency units available and in case of losses larger than this amount systemic risk would realize. Hence, we apply a threshold approach on the system level for defining systemic risk in this case. Let's further assume that due to his risk aversion he wants to be saved for all events up to the 100-year event. Given these assumptions, he wants to lower the losses up to this risk level through decreasing the dependency between the nodes. In doing so, he performs the simulation approach again but now using the independence copula (in this case, the copula parameter θ is set to 1). For the independent case, i.e., no dependency between the nodes, he would be able to decrease the system level losses of a 100-year event down to 19 which he would be therefore satisfied with. Note, he still has 1 currency unit available and can, for example, further use this amount to invest in risk reduction for one or both nodes (e.g., based on the risk-layer approach as will be discussed in the next section).

In our third example, we take a look at a more complex network structure and look at increases in systemic risk due to individual risk failures. The example is taken from Hochrainer-Stigler et al. (2018e). The system under consideration has seven individual nodes that can be at risk (we assume again only downside risk and therefore the distributions can be interpreted as loss distributions) and are modeled again as random variables that exhibit a Gamma distribution with mean 10 and a variance of 20 (as in the first example). The left-hand side of Fig. 3.11 is showing the network without any connections, while the right-hand side shows a fully connected network.

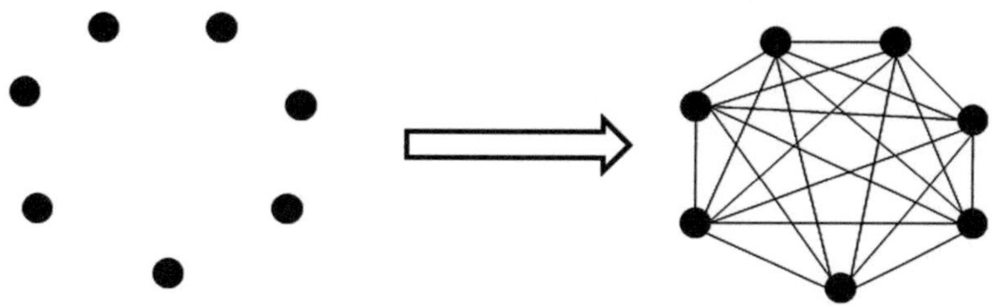

Fig. 3.11 Example of two dependency scenarios for a 7 node network example, no (left) and full (right) interactions. *Source* Based on Hochrainer-Stigler et al. (2018e)

For the dependency, a bivariate Joe copula between all nodes is assumed which should mimic tail dependence as discussed. Similar to example 2 and according to Fig. 3.11, we assume for the θ parameter of the bivariate Joe copula to be 1 (no dependence) or 8 (strong dependence) between all possible pairs. Based on the simulation approaches as discussed above, one is able to derive at a corresponding loss distribution. Given the availability of a loss distribution on the individual as well as system level, we now define extreme as well as systemic risks using the same measures, in our case we use the Value at Risk (VaR) at the $\alpha = 0.95$ level as risk measure for the individual and systemic risk. Hence, contrary to example 2 (which included a capacity estimate to deal with system-level losses) in this example we use risk measures to define systemic risk.

For the unconnected network, a VaR of 60 is found while for the strongly linked network the VaR is more than twice as large (around 129). Hence, if one would have assumed that a stress test should be performed on the system level and would have used the Value at Risk on the 95% level to determine the necessary backup capital needed and just looked at the unconnected cases, this would have lead to a severe underestimation of risk and capital requirements. Systemic risk, here defined as the risk that one has losses larger than the Value at Risk (on the 95% level), is much higher due to the dependence between the individual risks. Importantly, it is possible to determine the increase in capital requirements due to the linkages between the individual nodes, e.g., due to dependencies within the system. This is just the difference between the Value at Risk of the unconnected and connected case, i.e., 69 units. This number could be used as a measure of how much of the risk needs to be financed due to the connectedness of the individual elements within the system, i.e., due to systemic risk (see also Pflug and Pichler 2018 and the discussion further down below).

Equally if not even more important is the ability to estimate the influence of given nodes to systemic risk. For illustration, we repeat the calculations on the system-level risk based on conditional copulas as discussed in Sect. 3.2. We are especially interested if given some stress is exceeded in one node how the whole system would respond to this stress. Figure 3.12 indicates this situation by showing how dependency and systemic risk (defined here again through the Value at Risk) are connected

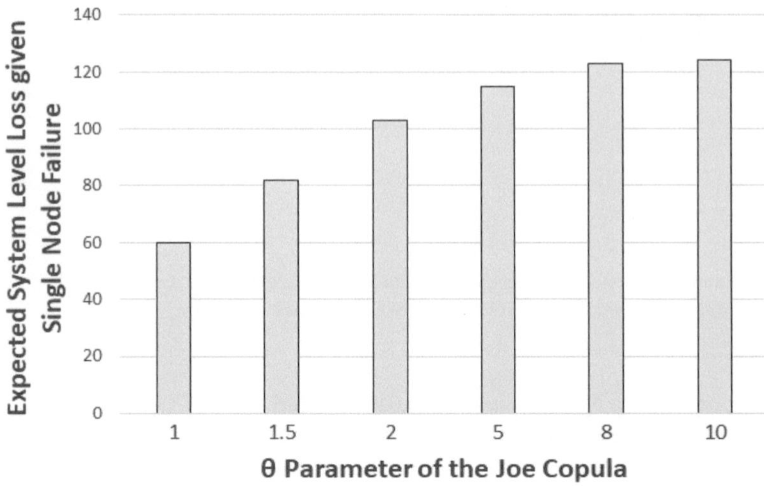

Fig. 3.12 Single node failure and expected system level losses. *Source* Based on Hochrainer-Stigler et al. (2018e)

contingent that there is already a failure within one single node of the seven-node network system. Again, single node failure is defined with the Value at Risk concept, i.e., losses are above the 95% Value at Risk, which (see Sect. 2.1) will happen with 5% probability. As systemic risk measure, we use the total expected system losses given that loss in one node is above the 95% level, defined as (see Hochrainer-Stigler et al. (2018e))

$$E_\theta\left(\sum_{i=2}^{7} L_i \mid L_1 \geq VaR_{0.95}\right). \tag{3.19}$$

Figure 3.12 indicates the importance of dependency for systemic risk as there is a steep increase in total losses given a single node failure even if only moderate dependency within the network is assumed. Hence, dependency of nodes within a network is quite important for the modeling of total losses from a systems perspective and especially for systemic risks that occur due to tail dependence of elements within the system.

From Fig. 3.12, it becomes clear that the copula parameters are not only helpful for systemic risk modeling purposes but can also be helpful as a possible warning indicator, i.e., that a system exhibits too much dependence between nodes. Obviously, in our network example above, one can either change the individual risks that will lead to systemic risks (e.g., change the Value at Risk to the 99% level, e.g., due to higher resources available) or one can change the dependency between the individual risks while keeping the individual risks the same. One extreme case of the later is to modularize or clusterize the system at hand, or in other words building subsystems. We make a comparison of different options as discussed in Hochrainer-Stigler et al.

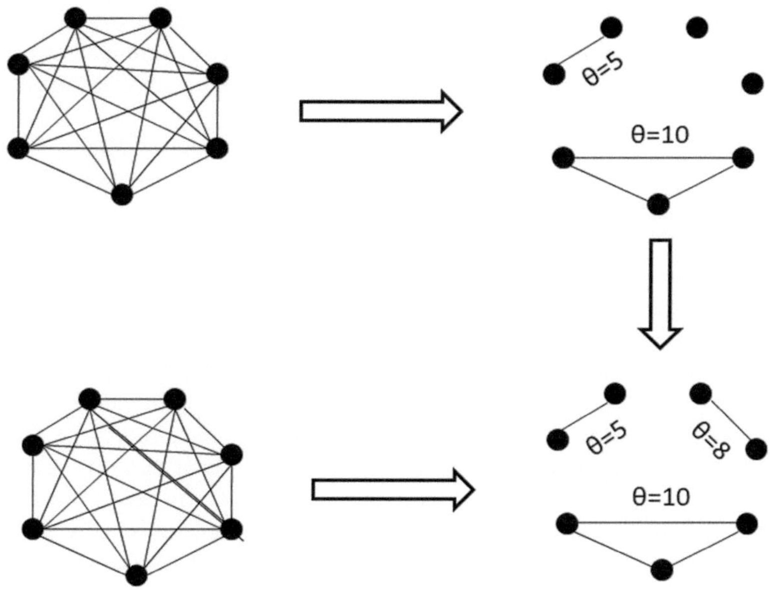

Fig. 3.13 Clustering of a 7 node network. *Source* Adapted from Hochrainer-Stigler et al. (2018e)

(2018e); however, focus here on the θ parameter of the Joe copula as exemplified in the formula given in Sect. 3.1.

As indicated, diversification strategies for the management of system-level risk can be applied in our approach either through modularization (e.g., building subsystems) or decrease of tail dependence, and Fig. 3.13 shows for illustration some possible ways to decrease risk. While risk diversification via modularization or decrease in interdependence between nodes can result in non-obvious changes in the risk on the system level, the suggested approach can fully quantify these changes. If one considers again the 95% Value at Risk for total losses as the measure for systemic risk and the fully independent network having a VaR of 90 and the strongly dependent having a VaR of 129, one can test on how modularization can decrease risk. In the case of decreasing the connected nodes as shown in the upper right-hand corner of Fig. 3.13, one can decrease the systemic risk by 28 points. Through an increase of a pair of nodes as shown on the lower right-hand side of Fig. 3.13, an increase of 2 in the systemic risk would be experienced. Hence, a copula approach for systemic risk may be especially useful in practical applications where some connectedness is desirable but system-level risk should not be above a pre-defined level. Furthermore, while we have looked here at just downside risk, the available copulas as well as distributions are, in principle, also able to distinguish between benefits and costs due to an increase in connectedness, e.g., the right tail of a distribution may show benefits while the left tail indicates losses. In that regard, there is also the possibility that there are more complex forms of dependencies within the individual risks possible which can balance potential system-level losses out, e.g., an increase in simultaneous loss events within a subsystem but a corresponding increase of simultaneous benefits within another (e.g., for crop yields in different breadbaskets around the world see Gaupp et al. 2017).

After these three examples which should have illustrated the usefulness and possible practical applications through this new perspective, we now want to discuss some additional advantages from a broader perspective. To start with, it became evident that two systems may look very similar in terms of the nodes and individual risk they are exposed too; yet, depending on the strengths of interdependencies the behavior of the system especially in times of crisis, including systemic risk realization, may be quite different (example 3). In addition, the conditional dependency that may trigger systemic risk may only occur in extreme cases such as visualized in Fig. 3.9, and consequently such dependencies cannot be observed during normal times but only realized during exceptional circumstances. This means that most of the times one will not know a lot about tail dependencies in a system if only past observations are taken into account. Data scarcity as discussed in Chap. 2 has to be assumed in many cases, and there is the need to establish approaches which at least sensitivity wise test the consequences of changes in assumed dependencies on the system level. As one possible way forward, one can start to change the shape parameters of the copulas and determine increases or decreases of systemic risk as well as the magnitude of systemic risks within this changed setting. For example, Borgomeo et al. (2015) used such an approach in the case of drought events in London by changing the copula parameters for the drought duration under climate change considerations.

It should also be noted that while the copula concept is essentially a statistical concept and therefore needs as a pre-requisite data to model the interdependencies, it can also be instrumental if the underlying drivers and interactions of elements within a system are yet not well known or understood but tail dependence has to be expected or must be assumed due to empirical observations or precautionary considerations. Phenomena of tail dependence are especially noticeable in hydrological processes such as water discharge levels across basins. As will be discussed, in the work of Jongman et al. (2014), it was shown that large-scale atmospheric processes can result in strongly correlated extreme discharges across river basins. The ability to physically model the flood phenomena on such large scales is not possible (due to data limitations and limited knowledge of the quality of river structures, e.g., sediments that affect water runoff); however, a copula approach circumvents this limitation by using empirical observations of past extreme discharge levels as a proxy for corresponding flood events. This enables new possibilities to model risk management strategies (see also Hochrainer-Stigler et al. 2018a; Hochrainer-Stigler et al. 2018c). As will be further discussed in the next chapter, new stress tests for flood financing instruments can be performed and probability levels when systemic risks may realize determined. A copula approach may also be beneficial to determine the underlying network of a system. As was discussed in Sect. 3.2, different spanning trees can be developed with such an approach which can give important indications how the elements within a system may be connected. Hence, while a copula can be applied to an already known network of interconnected agents, it is also useful for identifying a possible network structure given enough data is available.

Important in regard to our discussion on how to measure individual and systemic risks, Pflug and Pichler (2018) have suggested a measure of systemic risk which is in the spirit of some economic approaches about disaster events and indirect consequences (see Hochrainer 2009). The authors suggest to look at the difference in risk (for example, using the Value at Risk as a risk measure) on the system level between the independent and the interdependent case and to treat systemic risk as only the part of the risk which is due to interdependencies (while the other part of the risk is due to just individual risks). This idea can be related to our approach as well. Currently, systemic risk was defined through risk measures or thresholds on the system level. However, there is also the possibility to measure systemic risk using the proportion of individual-level failures as well. In more detail, given that systemic risk describes the propensity for cascading failures triggered by individual events, a continuum can be spanned by the proportion of individual failures with larger proportions characterizing risks at the more systemic end of the spectrum (Hochrainer-Stigler et al. 2018b). Hence, one may determine coping capacities on the individual level first and calculate the probability of default for each element separately. Afterwards, one calculates the probability of defaults under a tail-dependent assumption. The increase of the proportion of joint defaults is due to dependency and therefore would be part of systemic risk.

It should be also noted that up to now we have (explicitly or implicitly) assumed a decision-maker on the system level who is responsible to manage systemic risk. In reality, such decision-makers or institutions are not always set in place. As we have discussed, there are strong arguments to have a decision-maker who deals with risk from a system's perspective as increases in individual risk may be less significant for the risk of the whole system than increases in their interconnectedness (see also the discussion in Chap. 5). Already Helbing (2013) noted that before the financial crisis all the individual elements (e.g., banks, financial firms) operated optimal individually, while on the system-level systemic risk increased drastically but was not detected due to a lack of responsibility and accountability on that level. As we have seen during the financial crisis, major losses will likely be publicly borne and therefore it is in the interest of all to take such risks explicitly into account, e.g., by setting up respective institutions. We refer to Chap. 5 for a detailed discussion on this topic.

While the copula approach, as discussed here, is very beneficial in many regards, it also has limitations. For example, it may be of limited value if no suitable quantitative proxy for a dependent phenomena can be established (e.g., water discharges used as a proxy for large-scale flooding), it is also of limited value for exploring causal links and corresponding emergent behavior in complex systems. Our suggested approach is also of limited value if non-linearity within complex networks are the focal point of analysis. Other approaches are more suitable to address such issues, for example, Agent-Based Modeling (ABM) approaches. Additionally, while a copula can be used to determine the importance of nodes for systemic risk, they have limitations compared to measures such as DebtRank. Most obviously in a copula model, a node's impact on another node does not influence the impact on the same node as the impact ripples throughout the network, or in other words, no feedback loops are considered in our copula approach. The use of copulas is also not always appropriate, and other dependence measures may be used instead (Embrechts et al. 1999). Summarizing, while acknowledging the limitations of a copula approach for the integration of individual and systemic risk analysis, it still provides some entry points, conceptually as well as empirically, for many disciplines. Different scales of integration (local, regional, global) as will be discussed in the next chapter can be examined with different forms of copulas. There are also hybrid forms possible which are combining different approaches as well (see the application examples in Chap. 4), in the simplest case using copula-based results as an input to process-based approaches. What approach is most appropriate ultimately depends on the question, scale, data availability, and nature of the problem at hand. Regarding the management of individual and systemic risk, a comprehensive discussion based on resilience in combination of a so-called risk-layer approach is given next.

3.4 Resilience, Risk Layering, and Multiplex Networks

There are several reasons why individual as well as systemic risk should be analyzed in an integrated way (the following discussion is partly based on Hochrainer-Stigler et al. 2018b). Part of the motivation can be related to experiences in the past which have shown that there is a real potential nowadays for single events to trigger systemic risks. A prominent example is the 2011 Thailand floods that caused consequences globally (see Haraguchi and Lall 2015). Also, a review of globally relevant systemic risks by Centeno et al. (2015) found strong indications that individual events, even on the very local level, may cause large repercussions on the global scale. Additionally, the guidelines for systemic risk governance by the International Risk Governance Center (Florin et al. 2018) also placed strong emphasis on how individual failures may trigger systemic risks. In that regard, there have been recent calls for research on systemic risks to shift away from a component-oriented perspective to one that is interaction- and network-oriented (Helbing 2013). The approach discussed in this book also calls for an integrative perspective but, however, suggests to look at both aspects, the components and their interactions, simultaneously.

Up until now, the focus was on one network only; however, as we operate with loss distributions on both levels to determine individual and systemic risks, an integration over several levels or scales (so-called multiplex networks, here simply defined as network of networks) is also possible. In that sense, systemic risk realization on one level may be viewed only as an individual risk realization on another level. In other words, many components and their interactions at one level may be seen on a higher level as one individual component only which interacts with other components on that level creating systemic risk. Independent of the scale in one way or the other most of the time for systemic risk to realize, there must be at the very beginning a failure happening within one (or more) of the individual system components. Up to now, individual failure and systemic risk was defined through risk measures or

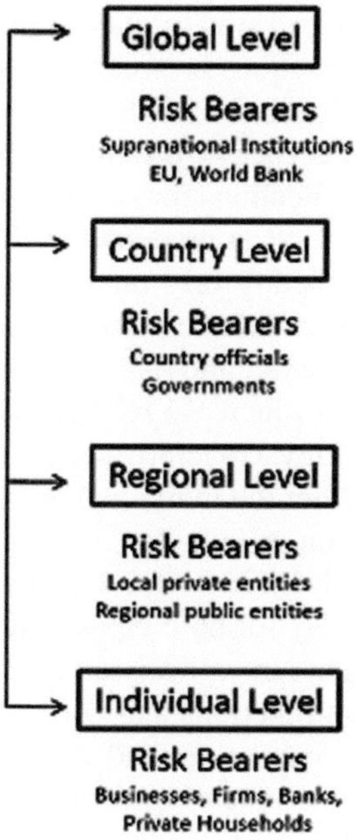

Fig. 3.14 Example of risk bearers across scales

thresholds, however, how a given individual risk may cascade over different scales will ultimately depend on the resilience of the agents involved on each scale of the system or system of systems (see, for example, risk bearers in Fig. 3.14 that are eventually affected due to extreme natural hazard events). Given the importance of the concept of resilience for individual and systemic risk, we discuss related concepts now in more detail.

Unfortunately, no commonly accepted definition exists yet for resilience and rather different schools of thoughts can be identified (e.g., National Research Council et al. 2012 and for an overview see Keating et al. 2014). Specific features of resilience that are suggested in the literature (see Linkov et al. 2016 for a comprehensive discussion and references) may include the critical functions of a system or thresholds which, if exceeded, perpetuate a regime shift or either focus on the recovery time from degraded system performance, while others emphasize the possibility for adaptive management through the anticipation of emerging risks and learning from past events (to name but a few). For a history of the term resilience and its use we refer to Alexander (2013). From the still quite prominent engineering context, resilience is related to the resistance of a system against disturbances and to the time it takes to get back to equilibrium. A major breakthrough of the concept resilience, however, came especially due to the work from Holling who focused on the "persistence of systems and their ability to absorb change and disturbance and still maintain the same relationships between populations or state variables" (Holling 1973, p. 14). The important new perspective Holling introduced was to understand resilience through a careful assessment of the dynamic interactions between a system's components. While the dynamics were especially emphasized by Holling, it must be noted that individual failure is often equally important. The two different perspectives, either on individual or systemic risk, have major implications for strategies aiming to increase resilience. For example, to increase resilience to moderate individual risks, for instance, car accidents, modern societies typically rely on some form of diversification strategy, most significantly through insurance (Geneva Association and Others 2010). Based on Holling's argument for systemic risks, in contrast, fundamentally different approaches need to be applied to increase resilience, for example, restructuring a network's connectivity or topology. Generally speaking, the identification of strategies to increase resilience on either the individual or system level is often related (as well as restricted) to the question of how they are measured. We already indicated possible measures for individual risk in Sect. 2.1, and Table 3.3 summarizes the most common risk measures used today for both types of risk.

In regard to risk management strategies and in respect to individual risk represented through loss distribution functions and corresponding measures (Table 3.3), a so-called risk-layer approach can be applied to identify different risk layers and the most appropriate options for increasing resilience (Benson 2012). In the case of natural disasters, risk layering involves identifying interventions based on hazard recurrences. The approach is based on the principle that varying risk management strategies adopted by different risk bearers (households, businesses, and public sector) are appropriate for distinct levels of risk on the grounds of cost efficiency and the availability of risk reduction and financing instruments (Linnerooth-Bayer and Hochrainer-Stigler 2015). Hence, risk measures applied for individual risk (such as those in Table 3.3) can be related to the most appropriate instruments to increase resilience through risk layering (see Fig. 1.2). For example, for low- to medium-loss events that happen relatively frequently, risk reduction is likely to be cost-effective in reducing burdens. Such events are usually assessed using the mean and median of reduced losses to estimate the effectiveness of risk reduction measures, for example, via cost-benefit analysis (Mechler 2016). As the costs of risk reduction often increase disproportionately with the severity of the consequences, other instruments need to be considered as well (Hochrainer-Stigler et al. 2018d). For this reason, it is generally advisable to use risk-financing instruments for lower probability/higher impact disasters that have debilitating consequences (catastrophes); these are usually measured through deviation and tail measures. Finally, as shown in the uppermost layer, there will be a point at which it is generally too costly to use even risk-financing instruments against very extreme risks, which are hence called "residual risk". It should be noted that such a risk-layer approach is especially useful if quantitative risk information is available but less appropriate if important dimensions such as loss of life should be considered as well. Additionally, there are many instruments available that can also be very cost-effective for high risk layers, e.g., such as early warning systems. In Chap. 4, a more detailed example of how the risk-layer approach should be applied is given.

Table 3.3 Selected individual and systemic risk measures

Individual risk measures	Informal explanation	Systemic risk measures	Informal explanation
Cumulative distribution function	Describes the probability that a random variable is less than or equal to a given value	Systemic expected shortfall	Measures the propensity of an institution to be undercapitalized when the whole system is undercapitalized
Location measure: expectation	The average value of a random variable over a large number of experiments	Measures in copula models	Systemic risk is defined as conditional VaR given that the systemic event occurs. The institutions/agents follow a copula dependency
Location measure: median	The median separates the higher and lower half of the distribution function	Systemic risk indices	An estimate of the amount of capital an institution/agent would need to raise in order to function normally given a systemic event
Dispersion measure: variance	Measures how far numbers of a random variable spread from the expectation	Set-valued approach	The systemic risk is the set of addition vectors which make the initial capital allocation acceptable
Tail measure: Value at Risk	A quantile measure for a random variable	Default impact	The cascade starts with given capitals and these are stepwise reduced by the cascade till the final stage, where no further failures are possible
Tail measure: conditional VaR	Expected losses given a specific value at risk level is exceeded	Contagion index	The expected default impact in a market stress scenario
Tail measure: tail index	Rate of speed of the tail for a given distribution function	DebtRank	Recursively account the impact of the distress of an initial node across the whole network, walks not allowed to visit the same edge twice

Source Based on Hochrainer-Stigler et al. 2020

For illustration purposes, we take a look at urban flood risk in Afghanistan (as done in Chap. 1). The loss distribution is taken from the Aqueduct global flood analyzer (Winsemius et al. 2016) which among only a few ones provides information on urban flood risk on the country level. In more detail, losses for nine different return periods are available. For example, a 10-year event loss, e.g., a loss that happens on average every 10 years, would cause losses of about 118 million USD, while a 250-year event would cause losses of about 270 million USD. If we assume as depicted in Figure 1.2 that the low risk layer is from the 1-till the 100-year event, the middle risk layer from the 100- up to the 500-year event, and the high risk layer for all events which happen less frequent than a 500-year event, one can determine the Average Annual Losses (AAL) for each risk layer (through integration of the respective parts of the loss distribution), which can in turn indicate costs to reduce risk for each layer. For example, for Afghanistan AAL for the three risk layers are found to be 35, 0.38, and 0.05 million USD. The risk estimates for each risk layer can be used in a subsequent step to determine costs for risk reduction. For example, if one uses overall estimates of cost-benefit ratios for risk reduction to be around 1:4 (according to Mechler 2016), i.e., investment of 1 USD will reduce losses of about 4 USD, risk reduction to decrease risk in the first layer would cost around 8.8 million USD annually. Corresponding calculations for the other risk layers can be done similarly; however, usually tail measures are more appropriate here. For example, for the middle risk layer, one can use the actuarial fair premium and multiply it by 20 as well as add a corresponding tail measure index (e.g., a 250-year event loss, around 270 million USD) to determine the costs for risk financing (see, for example, Jongman et al. 2014), which for Afghanistan would result in annual costs of about 277 million USD. This may be seen as expensive but in light of possible reduction of additional indirect costs, e.g., due to business interruption, could still be considered as one feasible option in the future (see also the discussion in Chap. 4). Summarizing, a risk-layer approach is useful to determine risk management strategies to decrease or finance individual risks.

As in the case of individual risks, measures of systemic risks (Table 3.3) have the potential to inform risk management strategies aiming to decrease risk, and diversification may be seen as a very promising strategy here too. However, no generally applicable simple results, as in the case of insurance for individual risk, have yet been found with regard to the impact on resilience. This is due to several reasons most of them related to the many different spreading mechanisms that are in principle possible (the discussion is based on Hochrainer-Stigler et al. 2018b). First of all, contrary to the insurance case, diversification may not only enable risk sharing and facilitate post-failure recovery, but can also multiply the number of pathways through which risks propagate. Furthermore, increasing modularity (the degree to which the nodes of a system can be decoupled into relatively discrete components) may decrease risks for most parts of the system, potentially at the expense of impeding resilience for the system as a whole. We already discussed that a copula approach could circumvent this problem if one includes not only a loss distribution but also a beneficial part to analyze possible trade-offs. Additionally, resilience may not go hand in hand with efficiency, for example, in road systems, Ganin et al. (2017)

noted that not all resilient networks are also efficient and vice versa and there are subtle differences that need to be accounted for. Last but not least, strategies to build resilience in the short term must take care to avoid so-called erosive strategies that lead to medium- and long-term negative impacts on development and well-being, for example, by the short-term over-exploitation of natural resources.

Therefore, instead of using a one-size-fits-all rule of thumb, the reshaping of a network topology is best based on an examination of the specific contribution of each node to systemic risk and various scenarios should be played through, at least conceptually, at best also quantitatively. The goal should be to identify nodes that are too big to fail, too interconnected to fail, and the so-called keystone nodes, which at times of failure cause large secondary effects or lead to a network's complete breakdown (see the still relevant discussion in Paine 1995). Restructuring a network can then be enabled by acting on those nodes. For example, Poledna and Thurner (2016) proposed a risk measure based on DebtRank that quantifies the marginal contribution of individual liabilities in financial networks to the overall systemic risk. They then use this measure to introduce an incentive for reducing systemic risks through taxes, which they show can lead to the full elimination of systemic risks in the systems considered. The resultant proposal of a systemic risk tax is a concrete measure that can increase individual and systemic resilience. In other cases, more broadly based governance approaches may be necessary (Florin et al. 2018), which in turn might require changes in human behavior or cultural norms. We discuss these issues in more detail in Chap. 5.

It should become clear that increasing a system's resilience differs among individual risks and systemic risks and also is dependent on the approach and measures used for representing risk. For individual risks, we argued in this section that a probability distribution and corresponding measures that can be related to risk layers and options can be established and used to increase resilience. Furthermore, for each risk layer, various market-based instruments may already exist, including insurance, which can be readily implemented (see the discussion of management options in Sect. 2.3 too). In contrast, systemic risk strategies are not as straightforward to structure conceptually, and different transformational approaches need to be developed and applied. Focus should be on the contributions of nodes to systemic risk and restructuring networks in order to reduce their contributions. The copula and structure of copulas in a network can give important insights here and can provide information about strategies to increase resilience against systemic risk. Note, resilience of system-level risk for the independent case can be increased by the same options as those discussed for individual risk. The difference between system-level risk for the independent case and any dependent system case is the additional risk due to interdependencies which should be measured and dealt with as described by systemic risk measures and corresponding options. For each resilience strategy considered, the consequences for both, the independent and the dependent case, can be analyzed and the appropriateness of the strategy quantified. This also enables an iterative approach, which is currently seen as appropriate for systems which exhibit high uncertainty, for example, future climate change impacts (SREX 2012).

Summarizing this chapter, with the loss distribution approach (using Sklar's theorem), various scales and risk layers can be looked at in an integrated way, depending on the scale the risk bearer is looking from, e.g., from a bottom-up or top-down perspective. If all risk bearers on each scale within a multiplex network would apply this approach, it is suggested that it would be most useful to decrease risks. It would also make it easier to see possible spreading mechanisms as the connections between risk bearers get more tangible. We discussed that systemic risk results from the interactions of individual risks. As a consequence, we argued that systemic risk cannot be measured by separately quantifying the contributing parts only. As in the case of individual risk, measures of systemic risk can also differ quite significantly. We introduced some association measures for dependency in the context of risk. However, there are also other measures available, some focus on the influence of single nodes on overall systemic risks, some focus on the position of a node within the network, while others emphasize the structure of the entire network. Many measures have recently been suggested for financial systemic risk in particular, not only because of its immense importance for society, but also because of the availability of high-precision and high-resolution data. One of the most important systemic risk measures in financial network analyses today was already introduced, DebtRank (Battiston et al. 2012). Based on the individual and systemic risk measures, strategies to enhance resilience either for one or both can be developed. As was already indicated, some of them can be used within a copula-based approach while some not. Also, some can be used within a hybrid approach with some parts or scales (in case of a multiplex networks) of the network dealt with different types of approaches, e.g., copulas applied on one scale and cascading effects modeled through an ABM modeling approach on another scale (see Chap. 4). Hence, this chapter should have laid out how loss distributions and Sklar's theorem can enable an integrated perspective on individual and systemic risk and resilience over multiple scales and agents.

References

Aas K, Czado C, Frigessi A, Bakken H (2009) Pair-copula constructions of multiple dependence. Insur: Math Econ 44(2):182-198

Alexander DE (2013) Resilience and disaster risk reduction: an etymological journey. Nat Hazards Earth Syst Sci 13(11):2707–2716

Baecker D (2016) Schlüsselwerke der Systemtheorie. Springer, Wiesbaden

Barabási A-L, Albert R (1999) Emergence of scaling in random networks. Science 286(5439):509–512

Battiston S, Puliga M, Kaushik R, Tasca P, Caldarelli G (2012) Debtrank: too central to fail? Financial networks, the fed and systemic risk. Sci Rep 2:541

Benson C, ArnoldM, de la Fuente A,Mearns R (2012) Financial innovations for social and climate resilience: establishing an evidence base. Social Resilience & Climate Change Brief, The World Bank, Washington, DC, USA

Borgomeo E, Pflug G, Hall JW, Hochrainer-Stigler S (2015) Assessing water resource system vulnerability to unprecedented hydrological drought using copulas to characterize drought duration and deficit. Water Resour Res 51(11):8927–8948

Centeno MA, Nag M, Patterson TS, Shaver A, Windawi AJ (2015) The emergence of global systemic risk. Ann Rev Soc 41:65–85

Clarke KA (2007) A simple distribution-free test for nonnested model selection. Polit Anal 15(3):347–363

Council NR et al (2012) Disaster resilience: a national imperative. Washington, DC: The National Academic Press

Czado C, Brechmann EC, Gruber L (2013) Selection of vine copulas. Copulae in mathematical and quantitative finance. Springer, Berlin, pp 17–37

Denuit M, Dhaene J, Goovaerts M, Kaas R (2006) Actuarial theory for dependent risks: measures, orders and models. Wiley, New York

Dissmann J, Brechmann EC, Czado C, Kurowicka D (2013) Selecting and estimating regular vine copulae and application to financial returns. Comput Stat Data Anal 59:52–69

Durante F, Sempi C (2015) Principles of copula theory. Chapman and Hall/CRC, Boca Raton

Elias N (1987) Involvement and detachment. Basil Blackwell, Oxford

Embrechts P, Mcneil E, Straumann D (1999) Correlation: pitfalls and alternatives. Risk Mag 12(5):69–71

Florin M-V, Nursimulu A, Trump B, Bejtullahu K, Pfeiffer S, Bresch DN, Asquith M, Linkov I, Merad M, Marshall J et al (2018) Guidelines for the governance of systemic risks. Technical report, ETH Zurich

Ganin AA, Kitsak M, Marchese D, Keisler JM, Seager T, Linkov I (2017) Resilience and efficiency in transportation networks. Sci Adv 3(12):e1701079

Gaupp F, Pflug G, Hochrainer-Stigler S, Hall J, Dadson S (2017) Dependency of crop production between global breadbaskets: a copula approach for the assessment of global and regional risk pools. Risk Anal 37(11):2212–2228

Genest C, Rivest L-P (1993) Statistical inference procedures for bivariate archimedean copulas. J Am Stat Assoc 88(423):1034–1043

Geneva Association and Others (2010) Systemic risk in insurance: an analysis of insurance and financial stability. Special report of the Geneva Association Systemic Risk Working Group, March

Haraguchi M, Lall U (2015) Flood risks and impacts: a case study of thailand's floods in 2011 and research questions for supply chain decision making. Int J Disaster Risk Reduct 14:256–272

Helbing D (2007) Managing complexity: insights, concepts, applications. Springer, Berlin

Helbing D (2013) Globally networked risks and how to respond. Nature 497(7447):51

Hochrainer S (2009) Assessing the macroeconomic impacts of natural disasters: are there any?. The World Bank, Washington, DC

Hochrainer-Stigler S (2016) Systemic and extreme risks: ways forward for a joint framework. IDRiM. In: 7th international conference on integrated disaster risk management disasters and development: towards a risk aware society, 1–3 October 2016. Islamic Republic of Iran, Isfahan, p 2016

Hochrainer-Stigler S, Lugeri N, Radziejewski M (2014a) Up-scaling of impact dependent loss distributions: a hybrid convolution approach for flood risk in europe. Nat Hazards 70(2):1437–1451

Hochrainer-Stigler S, Balkovič J, Silm K, Timonina-Farkas A (2018a) Large scale extreme risk assessment using copulas: an application to drought events under climate change for Austria. Comput Manag Sci 1–19

Hochrainer-Stigler S, Boza G, Colon C, Poledna S, Rovenskaya E, Dieckmann U (2018b) Resilience of systems to individual risk and systemic risk. In: Trump B et al. (eds) IRGC Resource Guide on Resilience, vol 2, Lausanne, CH

Hochrainer-Stigler S, Keating A, Handmer J, Ladds M (2018c) Government liabilities for disaster risk in industrialized countries: a case study of australia. Environ Hazards 17(5):418–435

Hochrainer-Stigler S, Linnerooth-Bayer J, Mochizuki J (2018d) Flood proofing low-income houses in India: an application of climate-sensitive probabilistic benefit-cost analysis. Econ Disasters Climate Change 1–16

Hochrainer-Stigler S, Pflug G, Dieckmann U, Rovenskaya E, Thurner S, Poledna S, Boza G, Linnerooth-Bayer J, Brännström Å (2018e) Integrating systemic risk and risk analysis using copulas. Int J Disaster Risk Sci 9(4):561–567
Hochrainer-Stigler S, Colon C, Boza G, Brännström Å, Linnerooth-Bayer J, Pflug G, Poledna S, Rovenskaya E, Dieckmann U (2019) Measuring, modeling, and managing systemic risk: the missing aspect of human agency. J Risk Res 1–17
Hochrainer-Stigler S, Colon C, Rovenskaya E, Poledna S, Boza G, Dieckmann U (2020) Enhancing Resilience of Systems to Individual and Systemic Risk: Steps toward An Integrative Framework (submitted)
Hofert M, Kojadinovic I, Mächler M, Yan J (2018) Elements of copula modeling with R. Springer, Cham
Holling CS (1973) Resilience and stability of ecological systems. Ann Rev Ecol Syst 4(1):1–23
Joe H (1997) Multivariate models and dependence concepts. Chapman and Hall/CRC, New York
Joe H (2014) Dependence modeling with copulas. Chapman and Hall/CRC, Boca Raton
Jongman B, Hochrainer-Stigler S, Feyen L, Aerts JC, Mechler R, Botzen WW, Bouwer LM, Pflug G, Rojas R, Ward PJ (2014) Increasing stress on disaster-risk finance due to large floods. Nat Climate Change 4(4):264
Keating A, Mechler R, Mochizuki J, Kunreuther H, Bayer J, Hanger S, McCallum I, See L, Williges K, Hochrainer-Stigler S et al (2014) Operationalizing resilience against natural disaster risk: opportunities, barriers, and a way forward. Zurich Flood Resilience Alliance
Klugman SA, Panjer HH, Willmot GE (2012) Loss models: from data to decisions, vol 715. Wiley, New York
Kovacevic RM, Pflug G (2015) Measuring systemic risk: structural approaches. In: Quantitative financial risk management: theory and practice, pp 1–21
Linkov I, Trump BD, Fox-Lent C (2016) Resilience: approaches to risk analysis and governance. In: An edited collection of authored pieces comparing, contrasting, and integrating risk and resilience with an emphasis on ways to measure resilience, EPFL International Risk Governance Center, Lausanne, p 6
Linnerooth-Bayer J, Hochrainer-Stigler S (2015) Financial instruments for disaster risk management and climate change adaptation. Clim Change 133(1):85–100
Lugeri N, Kundzewicz ZW, Genovese E, Hochrainer S, Radziejewski M (2010) River flood risk and adaptation in Europe–assessment of the present status. Mitigation and adaptation strategies for global change 15(7):621–639
Luhmann N (1995) Social systems. Stanford University Press, Stanford
Luhmann N, Baecker D, Gilgen P (2013) Introduction to systems theory. Polity, Cambridge
McNeil AJ, Frey R, Embrechts P (2015) Quantitative risk management: concepts, techniques and tools, Revised edn. Princeton University Press
Mechler R (2016) Reviewing estimates of the economic efficiency of disaster risk management: opportunities and limitations of using risk-based cost-benefit analysis. Nat Hazards 81(3):2121–2147
Nelsen RB (2007) An introduction to copulas. Springer Science & Business Media, New York
Paine RT (1995) A conversation on refining the concept of keystone species. Conserv Biol 9(4):962–964
Pflug GC, Pichler A (2018) Systemic risk and copula models. Cent Eur J Oper Res 26(2):465–483
Poledna S, Thurner S (2016) Elimination of systemic risk in financial networks by means of a systemic risk transaction tax. Quant Financ 16(10):1599–1613
Ruppert D, Matteson D (2015) Statistics and data analysis for financial engineering. Springer, New York
Sachs L (2012) Applied statistics: a handbook of techniques. Springer Science & Business Media
Sklar M (1959) Fonctions de repartition an dimensions et leurs marges. Publ Inst Stat Univ Paris 8:229–231
SREX (2012) Special report of working groups I and II of the intergovernmental panel on climate change (IPCC). Cambridge University Press

Thurner S, Hanel R, Klimek P (2018) Introduction to the theory of complex systems. Oxford University Press, Oxford

Timonina A, Hochrainer-Stigler S, Pflug G, Jongman B, Rojas R (2015) Structured coupling of probability loss distributions: assessing joint flood risk in multiple river basins. Risk Anal 35(11):2102–2119

Trivedi PK, Zimmer DM et al (2007) Copula modeling: an introduction for practitioners. Found Trends® Econ 1(1):1–111

Winsemius HC, Aerts JC, van Beek LP, Bierkens MF, Bouwman A, Jongman B, Kwadijk JC, Ligtvoet W, Lucas PL, Van Vuuren DP et al (2016) Global drivers of future river flood risk. Nat Clim Change 6(4):381

Chapter 4
Applications

In Chap. 2, we introduced individual risk defined as a random variable characterized by its distribution function. In case of pure downside risk, e.g., losses, we called such a distribution a loss distribution. We treated extreme risk estimation separately there as a theory of its own was needed for such kind of risks, namely, Extreme Value Theory (EVT). Afterwards, in Chap. 3 we defined a system as a set of interconnected individual elements which are “at risk” and introduced Sklar’s theorem (Sklar 1959) which provides a one-to-one relationship between multivariate distributions and so-called copulas. We suggested to reinterpret Sklar’s theorem from a network perspective, treating copulas as a network property and individual risk as elements within the network. In that way, we argued that both, individual and systemic risks, can be analyzed independently as well as jointly (Hochrainer-Stigler et al. 2018e). Importantly, on the individual as well as system level, a loss distribution can be established and we defined systemic risk similar as in the case of individual risk, i.e., through risk measures, e.g., Value at Risk, or through threshold assumptions (see Sect. 3.4). This approach can also be applied across several scales, or networks of networks, and consequently systemic risk realization on one level may be viewed only as individual risk realization on another level.

The ideas stated in Chap. 3 can be applied to various research domains, including financial, ecological, or natural disaster-related ones. The focus in this chapter will be limited to the later, i.e., natural disaster applications, as dependencies occur there very naturally (e.g., due to large-scale atmospheric patterns, or through interactions between different stakeholders during loss events). We already introduced in Sect. 2.2 an example how extreme value distributions for weather variables such as precipitation can be estimated. Such information can further be used to estimate a loss distribution either through attaching to each rainfall extreme a given loss (e.g., based on past observations) or using the loss observations themselves and applying Extreme Value Theory (EVT); the decision very often depends on data availability and application aspects, e.g., setting up indemnity-based insurance schemes or index-based insurance schemes (see in this context for the Malawi region Hess and Syroka 2005). Very often loss information from different local regions should be combined for risk analysis purposes and there are different ways how to move

© Springer Nature Singapore Pte Ltd. 2020

S. Hochrainer-Stigler, *Extreme and Systemic Risk Analysis*, Integrated Disaster Risk Management, https://doi.org/10.1007/978-981-15-2689-3_4

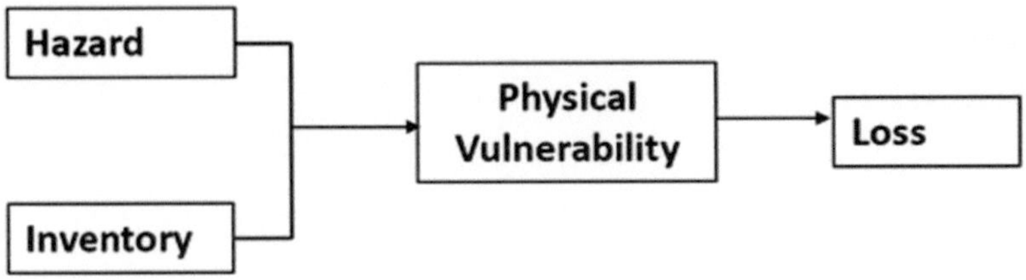

Fig. 4.1 Catastrophe modeling approach. *Source* Adapted from Grossi et al. (2005)

forward in that regard. One possibility is to apply a catastrophe modeling approach. The final goal, similar to EVT, of a catastrophe risk modeling approach is to derive at probabilistic loss estimates, e.g., due to natural hazard events. The standard approach for estimating natural disaster risk and potential impacts through a loss distribution is to understand natural disaster risk as a function of the natural hazard, the exposed elements, and its vulnerability (see for a discussion of these terms in more detail Birkmann 2006). Similar to this definition, Fig. 4.1 shows the catastrophe modeling approach in a nutshell (see for a comprehensive discussion Grossi et al. 2005).

Four components are usually needed to obtain probabilistic losses within a catastrophe modeling approach. The hazard component characterizes the hazard in a probabilistic manner. Very often all possible events which can impact the elements at risk are taken into account. These events are usually described by magnitude and associated annual probabilities, among other characteristics. Note, usually also events that haven't been observed yet are of interest and therefore EVT approaches are very often applied here too (Winsemius et al. 2016), see also the example of excess of rainfall as discussed in Sect. 2.2. The exposure component describes the inventory of assets such as single structures or a collection of structures that may be damaged due to an hazard event. The (physical) vulnerability module combines hazard intensities and damages to structures which are stored in the inventory component (Sadeghi et al. 2015). Very often physical vulnerability is characterized as a mean estimate of damage (e.g., average percentage of a house destroyed) given a hazard magnitude. Finally, the loss component combines the three components to estimate losses to the various stakeholders who must manage the risk (we refer for a detailed discussion to Grossi et al. 2005; Gordon 2011).

Catastrophe models usually produce probabilistic information, such as loss distributions, on the very local scale, e.g., over grid cells, i.e., regular shapes, with a given pixel resolution, say 250 m $\times$ 250 m. Usually, for each cell, the average annual losses are calculated and summed up to different geographical areas (e.g., political boundaries) (see Lugeri et al. 2010). However, while averages can be used as an indicator for the risk a region is exposed to, they do not capture the essence of extreme risks and as discussed in Sect. 3.4 risk instruments are difficult to be assessed. Especially for hazard events such as flooding or droughts, special considerations have to be taken into account as they are not spatial limited events and the dependency of a cell with other neighboring cells (respective regions) needs to be explicitly modeled for decision-makers on larger scales, e.g., regions, countries, or even on the global level (see Fig. 3.14). This is due to the fact that each decision-maker has different resources to cope with a disaster event, and ignoring dependent risks may severely underestimate losses and may cause a systemic event to realize as instruments which are set in place will likely fail exactly at that moment when they are most needed. Two specific forms of dependencies are looked at in this chapter: dependencies of hazards over regions and dependencies of stakeholders during natural disaster events. For example, according to our suggestions in Sect. 3.3, the grid cell loss distributions can be identified as individual risk, while the dependency between them (e.g., due to large atmospheric conditions, specific river structures, etc.) can be modeled through a copula approach. We will call the estimation procedure to derive at loss distributions on larger scales based on loss distributions on smaller scales that take the non-linear dependency explicitly into account as upscaling (Hochrainer-Stigler et al. 2014a). While we acknowledge the fact that the actual causal relationship for large-scale hazard events is due to complex interactions on the very local level, e.g., in the case of flooding river branches, sediments, etc., we want to emphasize that on larger levels it is very difficult to derive at probabilistic loss estimates and scenario approaches were mostly adopted in such kind of assessments in the past. This is not the case with the suggested approach as it can use proxies for tail dependencies which enables the integration of risk from the very local up to larger scales.

Regarding the dependency between different risk bearers (see Fig. 3.14), we want to give a short example first for illustration purposes. For instance, the risk of default of the European Union Solidarity Fund (which assists countries after large-scale disasters, see Sect. 4.1.2) is mostly due to dependencies of large-scale hazard events across countries, such as flooding (Jongman et al. 2014). A default would have consequences on the individual country level as less assistance would be available for governments. These governments are already at high stress as losses across basins are highly dependent too during such events and therefore would cause large losses and even more financial stress to the government (Hochrainer-Stigler and Pflug 2012). This will also affect risk bearers at the very local scale such as households, especially the poor, as they are usually in need of receiving assistance. From a household perspective, to increase resilience to hazard events one way forward could be to ask how systemic risk on larger levels that will ultimately impact the agents on lower levels can be reduced. This can be achieved either through strengthening individual resilience (e.g., through risk layering, see Sect. 4.2.1) or systemic risk resilience (e.g., through changes in the dependencies).

Based on the aforementioned discussion, this chapter is devoted to show in-depth applications of how to use the techniques presented in Chaps. 2 and 3. Furthermore, it will be shown how the dependencies on different scales for different risk bearers can be incorporated. Based on the discussions of techniques in Sects. 2.2 and 3.1, we look at three different copula and structural approaches focusing on countries in Europe and discuss how to upscale loss distributions and how they can be used for individual and systemic risk analyses. The focus will be mainly on floods and droughts. The first application in Sect. 4.1.1 will focus on a situation with very limited data availability. The second section (Sect. 4.1.2) focuses on a dynamic hydrological model and an advanced copula approach to derive at a loss distribution on the Pan-European level. The third section (Sect. 4.1.3) gives an example how to assess large-scale drought risk and present R-vines. The next section then focuses on measuring and risk management applications. The techniques and methods discussed in the previous chapters will be used extensively throughout this analysis. The case studies discussed were taken from various published literature on this subject (see the references in the respective sections).

4.1 Modeling Applications

As indicated, three different copula techniques to upscale individual risk to the system level will be presented in this section. All three vary in regard to their degree of complexity and scale. We start with the most simplistic one, move forward with a more advanced so-called minimax approach applied to flood events, and end this section with a discussion of a R-vine approach for large-scale droughts. All three applications use catastrophe modeling approaches as a starting point for estimating individual risk. Hence, they will be always discussed first.

4.1.1 Hierarchical Coupling: Flood Risks on the Regional Level

We first look at data and corresponding analysis from a study by Lugeri et al. (2010) which applied an overlay method for estimating direct flood risk. Regarding the hazard component, flood hazard maps were mainly employed in this approach and created from catchment characteristics. The central element used was a Digital Terrain Model (DTM) for the evaluation of the water depth in each location after a given flood extent. The specific flood hazard map used a 1 km $\times$ 1 km grid DTM and was combined with a dataset of European flow networks with the same grid size, developed at the Joint Research Center (JRC) (see Barredo et al. 2005). This provided quantitative information on both the expected extent of flooded areas and the water depth levels. It should be noted that the map was solely based on topographic features and an algorithm (developed at JRC) to compute the height difference between a specific grid cell and its closest neighboring grid cell containing a river (while still respecting the catchment tree structure). Therefore, the hazard map provided only "static" information which is quite different compared to hydrological model-based maps which provide probabilistic information arranged in terms of return periods (see, for example, Ward et al. 2013). As there was no probabilistic information on flood depths available at that time, a probabilistic read-out of the topography-based hazard map was performed through expert judgment and calibration from the LISFLOOD hydrological model (Van Der Knijff et al. 2010), which was available for

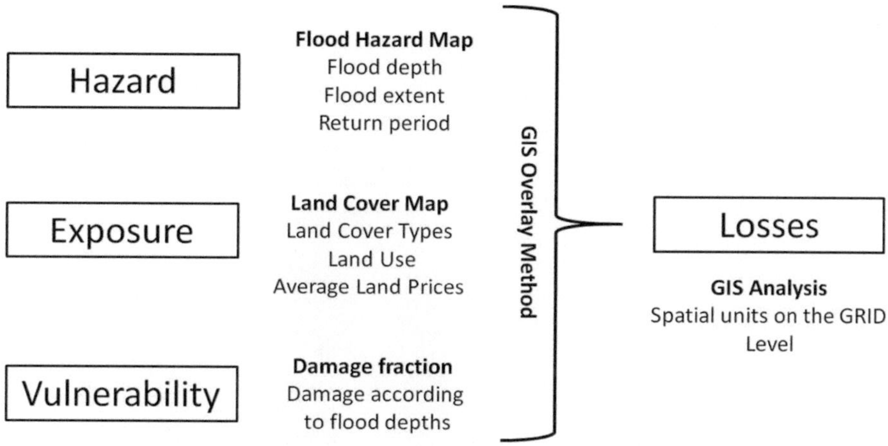

Fig. 4.2 Overlay approach for estimating flood damages. *Source* Based on Lugeri et al. (2010)

some catchments in Europe. The basic process was to attach to different water depth levels across Europe a corresponding event probability (Lugeri et al. 2010). As can be seen in Fig. 4.2, this approach is quite similar to the already introduced catastrophe modeling approach mentioned above.

The exposure component used the so-called CORINE (Coordination of Information on the Environment) Land Cover (CLC) map of the European natural and artificial landscape (see Feranec et al. 2016 for an introduction). The combination of the specific hazard maps described above with the availability of land cover maps allowed the estimation of the exposure to different flood events which in turn were grouped according to the different land classes of the CLC. While damage caused by floods depends on many characteristics of the flood itself (see Jongman et al. 2012), e.g., duration of the flood, water depth, or water flow velocity, given the scarcity of available data, the approach was the only way forward to combine hazard and exposure data. The last component, namely, the physical vulnerability of the assets under threat (in this case the land cover classes), was estimated by means of depth-damage functions for each land use class of the CLC which was available based on in-house JRC data. In a last step, the computation of monetary flood risk was done by GIS processing (this was basically the overlaying procedure) of hazard maps and the CLC and damage functions. The final outcome was grid-based maps (with pixel resolution of 250 m × 250 m) of monetary damages computed for five return periods (the 50-, 100-, 250-, 500-, and 1000-year return period).

Usually at this stage, one would calculate the average annual losses on the grid scale and upscale it through simple aggregation of the averages over specific regions, e.g., political boundaries (see Fig. 4.3). However, the probabilistic information would have been lost at larger scales, and therefore risk management approaches for individual (local level) and systemic (e.g., country level) risk would have been impossible to be applied. Furthermore, the simple assumption of independence between grid

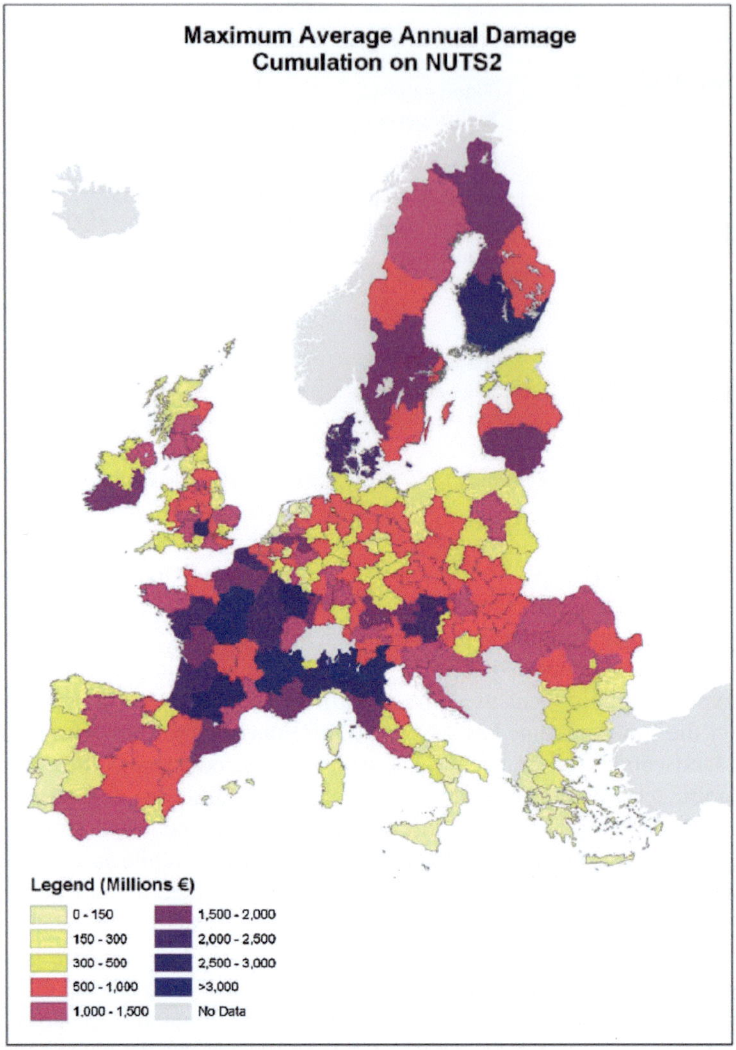

Fig. 4.3 Average annual flood losses on the Nuts 2 levels across Europe. *Source* Lugeri et al. (2007)

cells and corresponding loss distributions (see Sect. 3.1) would have underestimated the losses as large-scale flood events (e.g., during the summer in 2002 in Europe) can affect many regions at once. In other words, tail dependence has to be expected and should not be ignored. To avoid this underestimation, a copula approach based on hierarchical clustering using Strahler orders as well as a specific copula type that can distinguish between comonotonicity as well as independency across risk levels was used and will be discussed next (the following discussion is based on Hochrainer-Stigler et al. 2014a).

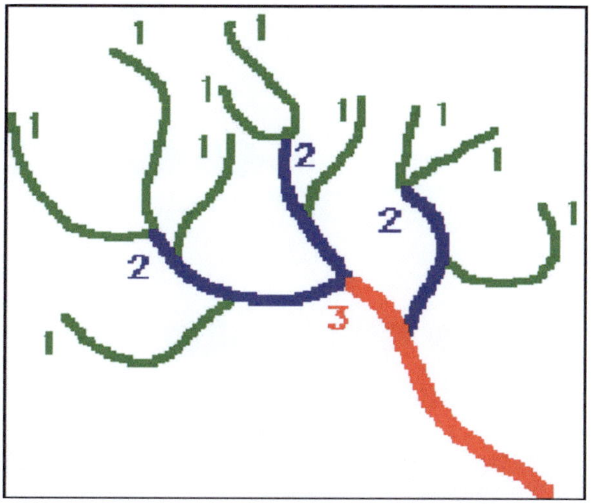

Fig. 4.4 A Strahler order example. *Source* Adapted from Hochrainer-Stigler et al. (2014a)

At the beginning, a specific pair-wise copula was introduced as due to data limitations it was not possible to estimate a copula from the Archimedean or other copula families. Therefore, it was assumed that up to a given probability level for two loss distributions X and Y, say $p*$ the random variables are assumed to be independent and beyond that point they are assumed to be co-monotone. The upscaled distribution of $Z = X + Y$ can then be given by separate formulae for each of the two parts, i.e., over the co-monotone part and over the independent part. This results in the following copula:

$$C(u, v) = \begin{cases} \frac{1}{p*}\min(u, p*) \, . \, \min(v, p*) & \text{if } \min(\text{u,v}) \leq p* \\ p * + (1 - p*) \, . \, \min(u, v) & \text{if } \min(\text{u,v}) > p* \end{cases}.$$

The estimation of the threshold probability $p*$ in the study by Lugeri et al. (2010) and Hochrainer-Stigler et al. (2014a) was based on simulation approaches for specific scenarios using the LISFLOOD model and therefore only gave some indications at what probability levels comonotonicity may be assumed. Note, at that time of the study, this was the only way forward and further down below much more sophisticated methods are presented. Nevertheless, the discussion here should demonstrate that even under very restricted data availability at least some ways for upscaling loss distributions are available which can be updated and improved once more data is available. Now, using the established pair-wise copula in the next step the structure how to combine the pair-wise copula over regions had to be determined. As flooding is very much dependent on the river structure, an approach based on Strahler orders was applied.

The basic idea was to start with a catchment and upscale distributions according to the river branches. When such catchments and corresponding loss distributions were upscaled, they were called clusters. The cluster levels were again then "marked"

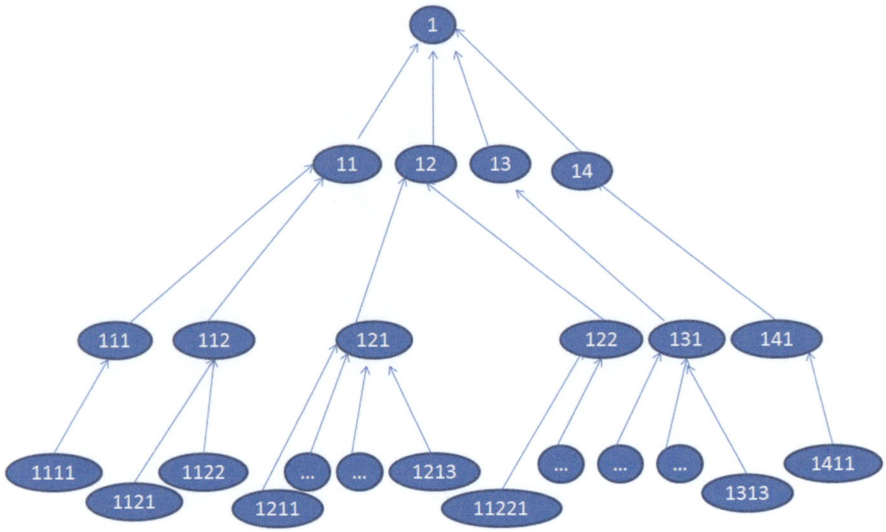

Fig. 4.5 Strahler order upscaling coding process

according to their "Strahler order". The term Strahler order (Strahler 1952) is a measure of the branching complexity of a network (applicable not only to rivers), which serves to classify the corresponding catchment. For example, starting from the lowest value 1, representing a river branch which has no tributaries, the Strahler order increases by one only when two (or more) branches with the same order meet (Fig. 4.4).

Consequently, Strahler order 1 defines a small catchment, Strahler 2 represents a river branch which upstream has at least two tributaries of order 1 (with a corresponding catchment which also includes the smaller ones relative to those order 1 tributaries), and so on. With this method, the cluster levels of each main river catchment can be identified. The clusters are treated separately until they are integrated according to the administrative boundaries. Figure 4.5 shows an example of how such a clustering scheme could be coded.

While the top cluster level 4 could represent a particular region, levels 3 to 1 are ordered in a way so that it is possible to determine which higher level cluster a given lower level cluster is part of. Note that in Fig. 4.5, cluster level 4 contains four level 3 clusters; each of them contains two, two, one, and one clusters of level 2, respectively, and so on, to keep track of the 13 elementary (level 1) clusters. The main river catchments in Europe show a higher level cluster code ranging from 6 to 9 cluster levels. After this hierarchical cluster definition process, the next task was to upscale the loss distributions according to the cluster level structure. Going from one cluster level (for example, cluster level $n-1$) to the next higher cluster level n, it is necessary to aggregate the loss distributions from all clusters in level $n-1$ within a given cluster level n. It should be noted again that by simple summation or convolution

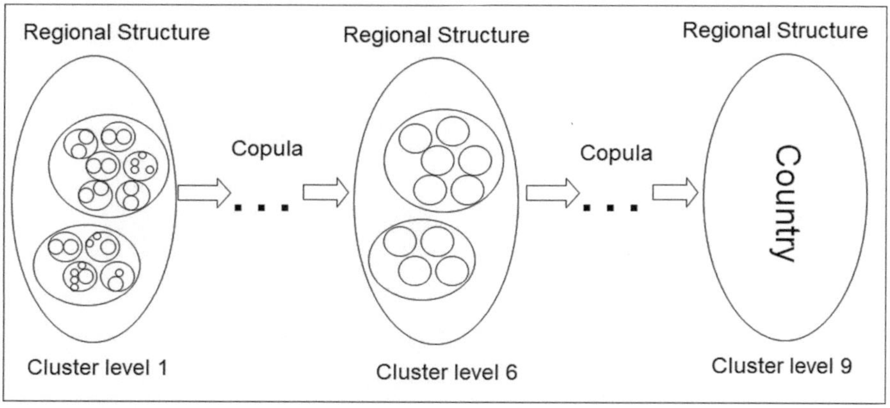

Fig. 4.6 Example of the upscaling process. *Source* Adapted from Hochrainer-Stigler et al. (2014a)

of all loss distributions for all clusters within a higher order level cluster would lead to over- or underestimation of risk (see also Sect. 4.2.1). This is due to the fact that by simple summation of the loss distributions it is implicitly assumed that losses of similar magnitude (respective risk) occurred at the same time everywhere in the region which would lead to overestimating the losses, while simple convolution (and therefore implicitly assuming independence) would underestimate the risk. Hence, for the aggregation of loss distributions to a higher cluster level, the aforementioned copula was used to upscale risk over all the relevant cluster levels (see Fig. 4.6).

The identified loss distributions on the grid scale as discussed above with the pair-wise copula and ordering approach explained was used to derive at loss distributions on the regional as well as country level. Here, we focus on Hungary, a country which is severely affected by large-scale flood events. Figure 4.7 shows the average annual losses (calculated through the area above the loss distribution, see Sect. 2.1) as well as the 100-year loss event for the case of the different regions of Hungary. One can see that losses are especially large in the Tisza regions located in eastern Hungary (including part of Central Hungary, Northern Hungary, the Northern Great Plain, and Southern Great Plain). Average losses are one indicator for risk; however, they do not capture extreme events and Fig. 4.7 also illustrates the 100-year loss for the respective regions in Hungary which again shows high risk in the Tisza regions where some regions can expect losses over 1 billion Euro for the 100-year event.

Also, the losses on the country level can be identified with a loss distribution through the upscaling method. For example, Average Annual Losses (AAL) on the country level are estimated to be around 128 million Euros and 100-year event loss would be around 5.7 billion Euros. Hence, using the copula and hierarchical upscaling procedure, the loss distributions on different system levels can be incorporated. See Table 4.1 for five selected annual return periods and corresponding losses for the seven main regions in Hungary. As can be seen, the losses differ quite dramatically between regions and not only in regard to more frequent losses but also very rare

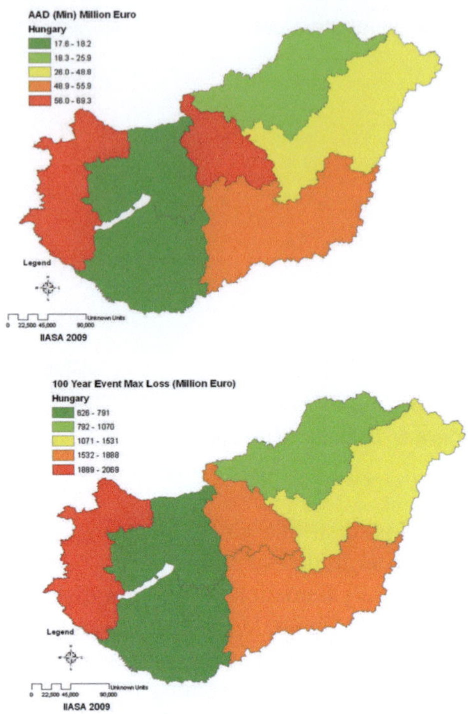

Fig. 4.7 Average annual losses for regions in Hungary and 100 year event losses. *Source* Figures based on data from Hochrainer-Stigler and Pflug (2012)

disaster events. For example, while compared to Western Transdanubia Central Hungary has lower losses for the year events below the 500-year event, it has considerable larger losses for the very extreme 1000-year event.

Such considerations may play an essential role for risk management purposes as will be discussed in one of the later sections.

4.1.2 *Minimax and Structural Coupling: Pan-European Flooding*

The previous approach used for upscaling loss distributions to the country level, exemplified for Hungary, had several limitations. First of all, no hydrological models and extreme value statistics were used in the process to determine the return periods for flood events on the grid scale. Furthermore, the selected copula and corresponding hierarchical structure were very simple and based on expert judgment and Strahler orders only. Finally, the inclusion of climate change and global change impacts for future risk assessment (see Kundzewicz et al. 2010) can only be done extremely simplistic, e.g., by moving the loss distribution according to changes in the 100-year return period (e.g., based on estimates from Hirabayashi et al. 2008). In this

Table 4.1 Regional probabilistic losses (in million Euros) for regions in Hungary

Region \Return period	50	100	250	500	1000
Central Hungary	806	1365	2835	3891	17333
Central Transdanubia	482	629	792	909	957
Western Transdanubia	885	1582	4073	7061	10984
Southern Transdanubia	351	476	745	1299	2397
Northern Hungary	617	861	1205	1580	1727
Northern Great Plain	885	1249	1986	3254	8466
Southern Great Plain	846	1504	2371	6861	7626

Source Based on data from Hochrainer-Stigler and Pflug (2012)

section, we present a more sophisticated model that was built in Jongman et al. (2014). We again want to discuss first the catastrophe modeling approach, how the loss distributions were upscaled through a copula approach, and afterwards want to discuss how they can be used for individual and systemic risk analysis. Special emphasis is given on how in the case of non-stationarity (see Sect. 2.2) one can use such modeling approaches to include climate change impacts. The focus in this section will be one scale higher than the country and regional levels discussed before and look at flood risk on the Pan-European level.

The hazard component which included climate change impacts as well was based on an ensemble of high-resolution (grid scales with pixel resolution of about 25 km $\times$ 25 km) climate simulations for Europe. In more detail, a combination of four Global Circulation Models (GCMs) and seven Regional Climate Models (RCMs) that covered the period from 1961 to 2100 at a daily time step and forced by the SRES A1B scenario was used. This is an old climate change scenario approach and nowadays rather the Regional Concentration Pathways (RCP) and Shared Socioeconomic Pathways (SSP) are looked at instead (see O'Neill et al. 2014). The A1B scenario is a subscenario within the A1 scenario and is defined in IPCC as follows.

"The A1 storyline and scenario family describes a future world of very rapid economic growth, global population that peaks in mid-century and declines thereafter, and the rapid introduction of new and more efficient technologies. Major underlying themes are convergence among regions, capacity building, and increased cultural and social interactions, with a substantial reduction in regional differences in per capita income. The A1 scenario family develops into three groups that describe alternative directions of technological change in the energy system. The three A1 groups are distinguished by their technological emphasis: fossil intensive (A1FI), non-fossil energy sources (A1T), or a balance across all sources (A1B)." (Nakicenovic et al. 2000).

Different to the analysis in Sect. 4.1.1 this time a large-scale hydrological model was available that could simulate the spatial-temporal patterns of catchment responses as a function of several variables including spatial information on meteorology, topography, soils, and land cover. Regarding land use properties, different soil types, vegetation types, land uses, and river channel information were available too. Furthermore, data on precipitation, air temperature, potential evapotranspiration, and evaporation from water bodies and bare soil were used as the main meteorological drivers. The so-called LISFLOOD model already mentioned now considered also processes such as infiltration, water consumption by plants, snowmelt, freezing of soils, surface runoff, and groundwater storage. The information from the high-resolution climate simulations was used as an input to the 5 km × 5 km grid resolution for LISFLOOD which simulated water volumes along river channels as primary output with a daily time step for the period 1961–2100. The model also provided river water levels (relative to channel bottom) estimated from the simulated water volumes and the cross-sectional (wetted) channel area of the river section. Extreme value analysis was employed to obtain discharge and water levels for every river pixel associated with different return periods (2, 5, 10, 20, 50, 100, 250, and 500 years). In more detail, a Gumbel distribution (see Sect. 2.2) was fitted to the 30 annual maxima values defined within four time windows (1961–1990, 1981–2010, 2011–2040, and 2041–2070), which were interpolated into a continuous series. This should account for non-stationarity effects due to climate change as one could assume at least weak stationarity for the respective time periods (see Sect. 2.2). For each of these time windows, eight return periods (2, 5, 10, 25, 50, 100, 250, and 500) were estimated and loss distributions were therefore available on the grid scale based on a dynamic and highly complex flood modeling approach.

The exposure component within this catastrophe modeling approach was again based on the land use classification of CORINE land cover (2006 version) but now country-specific depth-damage functions for the different land use classes were applied. To account for future exposure but acknowledging the fact that consistent land use projections were not available, it was assumed that the spatial distribution of the exposed elements at risk remained the same; however, a scaling factor based on future GDP projections from the A1B scenario was used to account for changes in exposure. Furthermore, no changes in the vulnerability depth-damage functions were assumed either (for a discussion of this dimension, see Mechler and Bouwer 2015). Usually, flood models take only the topography maps into account and do not incorporate protection standards implemented in different regions and hence such models overestimate losses. To account for this and also different to the first approach presented above in Jongman et al. (2014), protection standards were also incorporated in the analysis. In more detail, flood protection standards were defined as the minimum probability discharge that leads to flooding; in other words below the given discharge level, no losses are assumed to occur. The protection standards were determined through the analysis of policies implemented in major European river watersheds. The final protection standards ranged from 10-year return periods up to the 500-year return periods. It can be seen as the risk protection layer for frequent events within a risk-layer approach as discussed in Sect. 3.4.

As flood events are not spatial limited (as exemplified during the flood events in 2002 in Europe), the dependence between each cell (or catchments) with other cells (or catchments) had to be again explicitly taken into account (e.g., as this is very important for decision-makers on larger scales such as the European level). The aforementioned copula approach exemplified for the Hungarian case was exchanged with a more advanced technique based on a minimax approach as discussed in Timonina et al. (2015b) and which will be introduced next. The main reason for using a copula approach here was that while LISFLOOD provided loss distributions through a dynamic hydrological model it, however, is not able to account for the dependencies between the grid cells. One advantage in this study compared to the previous case was that time series of river discharges were available as well and used as a proxy for the dependency between loss distributions over larger regions. Also, Archimedean and Extreme value copulas could be estimated therefore too.

In a first step, the validity of using river discharges as a proxy for dependence was needed to be tested. As was shown in Timonina et al. (2015b), strong non-linear dependencies of river discharge levels within the Danube basins in Romania have to be assumed (see Fig. 1.5) and can be modeled through a flipped Clayton copula.

While the flipped Clayton copula empirically well described the flood loss behavior across different subregions (basins) of the Danube River Basin, also other Archimedean copulas were tested and used for the upscaling process as well (including a comparison with Gumbel and Frank copulas). However, the flipped Clayton copula was instrumental in the further analysis as it showed the best fit in most of the cases. Therefore, for each pair of basins (i, j), a flipped Clayton copula $C_{\theta_{ij}}$ was estimated by using maximum-likelihood estimation techniques as discussed in Sect. 3.1. In the next step, the θ parameters of the flipped Clayton copula between different pairs of basins were used within a minimax approach to determine the structure of how to upscale the loss distributions. In doing so, a set of N regions (e.g., basins) was assumed and a matrix with pair-wise estimates (based on the discharge levels) of the flipped Clayton copula was constructed, i.e.,

$$\Theta = \begin{pmatrix} 1 & \theta_{12} & \cdots & \theta_{1N} \\ \theta_{21} & 1 & \cdots & \theta_{2N} \\ \vdots & \vdots & \ddots & \vdots \\ \theta_{N1} & \theta_{N2} & \cdots & 1 \end{pmatrix}. \tag{4.1}$$

It is worth to be noted that the matrix Θ (note that the diagonal elements are not used) did not contain any information about the structure/topology of the geography of the region (for example, river structures over given regions), though interdependencies were found to follow some structure, especially the river topology (Fig. 4.8). It is also worthwhile to note that geographical nearness was not necessarily a good predictor of interdependencies between basins and rather the topology and the matrix served as a much better proxy. The pair-wise basins were upscaled using a so-called minimax approach which is closely related to ordered coupling techniques. The following algorithm was suggested in Timonina et al. (2015b) and is presented next. The algorithm for ordering vector θ_o out of matrix Θ started with the selection of the maximal θ_{ij} from the whole matrix Θ. This is similar to R-vine initialization approaches as discussed in Sect. 3.2, i.e., elements (i, j) that are most dependent in the tails are selected first. The ordering at this stage is $\theta_o = [i,j] = [j,i]$. In the next step, the element (i.e., region) k that is suitable for

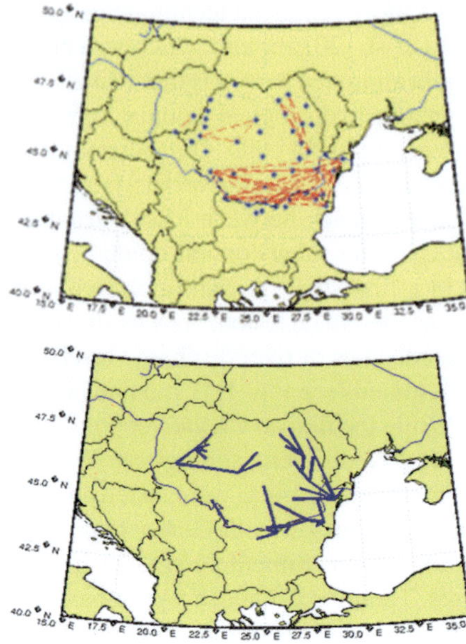

Fig. 4.8 Correlation between basins (above) and river structure of basins (below). *Source* Timonina et al. (2015a)

both regions i and j (so that it is dependent not only on the loss situation at region i, but also on the loss situation at region j) (notice that $k \neq i, k \neq j$) is selected. At this iteration, the reason why the approach is called minimax becomes apparent. One chooses the maximum of all minimum dependencies of all pair-wise copulas between the elements in θ_o and all other elements in Θ. In more detail, assume two vectors θ_l and θ_m: θ_l is a vector that contains dependencies between region i and all other regions 1,...,N and θ_m is a vector that contains dependencies between region j and all other regions 1,...,N. One constructs a new vector θ_n, such that its elements are minimal between θ_l and θ_m. Afterward, one maximizes over elements of vector θ_n and will get the element (or region) with index k. Hence, one derives at a triplet of regions $\theta_1 = [i, j, k]$, and the iteration is continued until the end of the vector with length N.

As indicated in the algorithm above, the measure for the interdependency between regions i and j was the flipped Clayton copula parameter θ_{ij}, i.e., refer to strong tail dependence if θ_{ij} is large and to independence in case of a small θ_{ij}. However, there is also the possibility to use also the simple Pearson correlation for the minimax structuring or just the geographical distance. Especially for the later, it was found that it is inadequate to be used for the copula structuring process (see Timonina et al. 2015b).

The copula approach as explained above was implemented to derive at a loss distribution on the Pan-European level. Furthermore, due to the information of future losses on the local level using the A1B scenario, it was also possible to include climate change impacts in the analysis. Summarizing the methodology, the minimax approach was applied to time-series data on maximal discharges for each basin in Europe for the period of 1990–2011. This served as a proxy to upscale the loss distributions available on the basin level to larger levels. The basin level loss distributions (interpreted as individual risk here) were derived using the LISFLOOD hydrological model and an economic damage model (Rojas et al. 2012). Furthermore, the geographical position of all basins (1433 in total) including longitude and latitude was available and used for upscaling the loss distributions according to specific boundaries (e.g., political ones). It should be noted that for the estimation of future loss distributions it was assumed that dependencies in the future stay the same as today. This may not be the case and would need some further analysis (see Borgomeo et al. 2015 for a kind of sensitivity analysis of risk due to changes in the copula structure). Finally, with this approach, loss distributions for today and in the future on the Pan-European scale were calculated. For example, it was found that large-scale flood losses such as the one in Europe in 2013 (around 12 billion Euro losses) have to be expected once every 16 years, on average, and its frequency will increase till 2050 to a 10-year event (Jongman et al. 2014). Furthermore, comparing the loss distributions with the independence assumption, it was found that losses on the EU level are magnitudes higher if tail dependence is incorporated. For example, a 100-year flood event on the Pan-European level would cause under the independent assumption around 10 billion losses while under the tail dependence assumption it would cause around 70 billion Euros of losses. Hence, a serious underestimation of risk has to be assumed if interdependencies are not accounted for. This should be seen also in relation to the discussion in Sect. 3.3 where it was noted that the system may behave quite differently under stress than under normal circumstances. The importance of dependency also raises the question if clustering and decoupling as a possible risk management approach may be an appropriate option as one can see that the independent assumption and corresponding estimated loss distribution is several magnitudes lower compared to the dependent case. Again having a loss distribution available different measures of risk can be applied to see which kind of risks would lead to a possible systemic event and we will focus in a later section on a special fund created on the EU level to partly assist countries under such large-scale events.

4.1.3 Vine Coupling: Large-Scale Drought Risk

Hierarchical as well as ordered coupling approaches were introduced in the two application examples above for flood risk. In the next application, we look at a different natural hazard that can happen over large geographical areas simultaneously, namely, drought. Here, we present the third introduced method within Sect. 3.2, i.e., applying an advanced R-vine approach including a selection of different copula families to upscale local loss distributions to higher levels. The discussion follows the methodology given in Hochrainer-Stigler et al. (2018a) which was applied again on the country level now for the case of Austria. It was already discussed in the case of floods that the risk-dependent nature of hazard events need to be taken into account for their assessment. As it was argued, due to atmospheric conditions, some hazards, including floods and droughts, are not always just local spatially limited phenomena but can affect large regions at once. This phenomenon was already discussed in the previous section but there are also other past events such as the heat waves in 2003 in Europe (or in Austria in 2015) which can serve as prominent examples here (see Stahl et al. 2016). As in the case of flooding also in the case of drought risk the neglection of tail dependencies may lead to a serious underestimation of risk, and therefore also to the wrong selection of management options (including funding needs). Also similar to flood risks, the past increase in worldwide losses due to drought events is alarming for many stakeholders including farmers, businesses, insurers, and governments and already today droughts pose a significant challenge to them (UNISDR 2009). Worryingly, the current extreme events which are happening are expected to worsen due to climate change in the future; however, the actual changes are difficult to be projected and large uncertainties exist (Field et al. 2012). Hence, it is important to investigate ways how to assess and manage drought risk for today and in the future.

To start with, it should be noted that there are different ways how to define a drought event. Usually, four different types of droughts are distinguished in the literature: meteorological drought, hydrological drought, agricultural drought, and socioeconomic drought (UNISDR 2009). As we are especially interested in crop losses, we focus on agricultural drought (hereinafter referred to simply as drought). As in the case of flood risk, the goal is to derive at loss distributions on different scales. This should be useful for determining as well as assessing different options to decrease current and future individual and systemic risks. According to Hochrainer-Stigler et al. (2018a), four challenges have to be addressed for large-scale drought risk assessment. The first challenge is about the estimation of extremes on the very local level. The second challenge is to account for non-stationarity issues due to climate change. The third challenge is to explicitly include tail dependence, and the fourth challenge is to upscale the risk to specific regional levels. As pointed out in Hochrainer-Stigler et al. (2018a), challenge one can be tackled through EVT, and challenge two can be tackled via biophysical crop growth models to simulate current and future crop yields under non-stationarities due to climate change effects (similar to the approach in Sect. 4.1.2 but now with an agricultural model based on Balkovič et al. 2013). The third and fourth challenges can be tackled simultaneously via a copula and R-vine approach. The overall methodology is shown in Fig. 4.9.

The basic input as in the case of catastrophe modeling approaches for the upscaling procedure is the loss distributions on the very local level. Losses in the case of drought are not asset losses but decreases in crop production. Hence, we are now interested in the left-hand side of a crop distribution which should indicate extreme events in regard to low crop yields (e.g., in comparison to average crop yields). To estimate crop growth and yield, usually biophysical crop models are employed. Main differences between such models are related to the different biophysical approaches they are using, e.g., representation of soil and crop processes and interventions (for a comparison of modeling approaches, see Rosenzweig et al. 2014). In the study of Hochrainer-Stigler et al. (2018a), the Environmental Policy Integrated Climate (EPIC) biophysical crop model was used as it can provide crop yield time

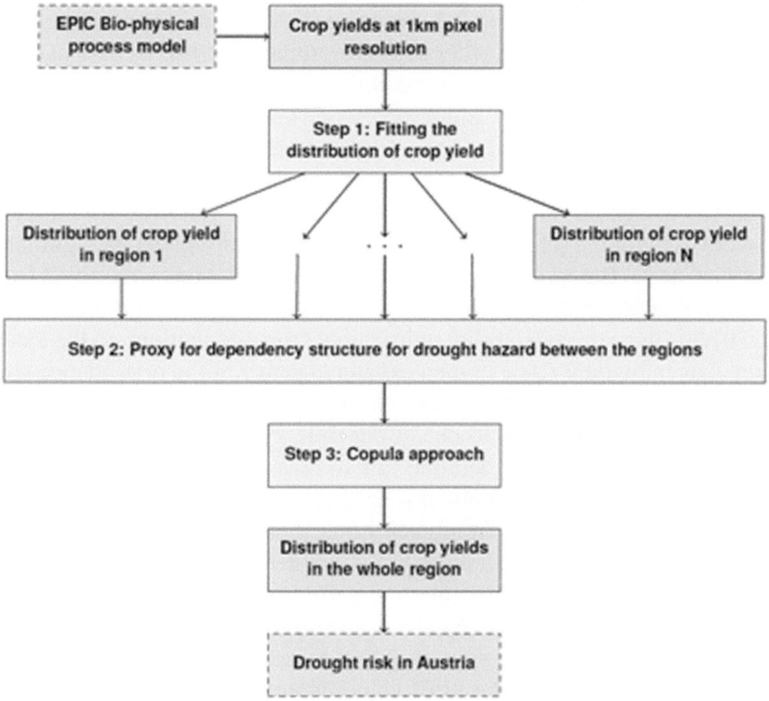

Fig. 4.9 Drought modeling approach. *Source* Based on Hochrainer-Stigler et al. 2018a

series for today and future scenarios. This information can be used to tackle challenge one, i.e., estimation of extreme events using EVT approaches. EPIC contains many routines for the simulation of crop growth and yield formation, hydrological, nutrient and carbon cycling, soil temperature and moisture, soil erosion, tillage, fertilization, irrigation and plant environment control, and so on. One key feature in EPIC is to derive at crop yields based on the actual above ground biomass at the time of harvest as defined by the potential and water-constrained harvest index. As in the other two cases above the main reason for using EPIC was to get gridded time-series data of annual crop yields simulated with a 1 × 1-km spatial resolution. For the future climate, an ensemble of bias-corrected climate projections was used. The data came from the so-called EURO-CORDEX (European branch of the Coordinated Regional Downscaling Experiment program) which was developed in the IMPACT2C project (Quantifying Projected Impacts Under 2 °C Warming project) using the RCP 4.5 scenario (see Van Vuuren et al. 2011).

Having a time series of crop yields available for today and the future in the next step a parametric approach for estimating the risk of low yields was used. In other words, crop yield distributions were fitted based on the simulation results of the EPIC model for today and future scenarios. As discussed in Chap. 2, this included the selection of various continuous distributions and the estimation of its parameters using the techniques discussed in Sects. 2.1 and 2.2, for example, applying Maximum-Likelihood (ML) estimation techniques. In order to find a proper (in terms of best fit) marginal distribution that could represent crop yield risk (for us the individual risk), various families as discussed in Sect. 2.1 and the GEV and GPD were used. After parameters for each SimU (Simulation Units, respective grid cells that cover the country) were estimated, the most appropriate distribution was selected based on Goodness of Fit (GoF) tests as discussed in Sect. 2.1 and 3.2. Finally, the AIC and BIC information criteria were used for the selection of marginal distributions, given the significance of the other GoF tests. Based on this approach, crop distributions were fitted for today (using the time series from 1971 to 2010) and the future 2050 (using time series from 2041 to 2070). Given the availability of marginal crop distributions on the local scale (interpreted as individual risk), in the next step a proxy of the dependence structure during drought events between regions had to be established.

As in the case of flood risk to detect dependencies between different SimUs a proxy for drought events is needed due to two reasons. First, no empirical data on crop yields is available on this granular scale and second, the simulations from EPIC in each SimU are independent from each other (hence, cannot be used as a proxy for dependence itself). Several indices are suggested in the literature for droughts. A very much accepted drought risk proxy was used here (playing a similar role as the discharge data for flooding), namely, the Standard Precipitation Evapotranspiration Index (SPEI, see Vicente-Serrano et al. 2010) to estimate the vulnerability of crop yields due to drought events. The SPEI values were calculated for every SimU and climate scenario with monthly time steps. The dry years are identified as years with the mean SPEI (calculated from Corn sowing to maturity) of less than -1, while normal years are defined as SPEI greater or equal -1. The SPEI index within a copula model can be used to determine the spatial dependence of drought events and has shown to reproduce past drought events well (Hochrainer-Stigler et al. 2018a, especially Supplementary II). However, there is another complication using the proxy which needs to be explicitly considered within the drought risk assessment. Due to irrigation as well as other management practices (similar to the case of protection levels for flood risk), one additionally needs to test if low SPEI values actually lead to low crop yields. As said, due to several reasons, e.g., irrigation or soil conditions, not all SimUs necessarily need to show dependency in crop yields during drought events as determined by the SPEI index. To account for these situations before the upscaling only those SimUs were selected which showed a significant dependency between low crop yields during droughts in the past (in terms of low average yields during low SPEI events). Hence, some additional indices had to be created to guarantee the selection of SimUs, which are interdependent in both the SPEI index and corresponding crop yields. The final selection of SimUs was based on the decision if the average crop yield during a long-term dry climate was lower than the expected value. SPEI data was based on historical meteorological data from JRC's Gridded Agro-Meteorological Data in Europe database with a 25-km grid resolution and time period from 1980 to 2015. The SPEI was spatially linked to the SimU-based crop yield simulations from the EPIC model. Having established a proxy for spatial and crop yield dependence, an R-vine approach was adopted as explained in Sect. 3.1.

The upscaling of the crop yields to the country level was performed in a two-step process. The historical SPEI data was separated into two parts, the ones which showed a relationship with crop yields during droughts and the ones which did not. Afterwards, both datasets were transformed into pseudo-observations in the unit square [0, 1] (see Sect. 3.1). Then, the tree structure using an R-vine approach for the dataset of the SPEI dependent cells was applied using the SPEI index as a proxy for dependency and using a maximum spanning tree approach (see Sect. 3.1). Finally, by applying the inverse transformation method, the R-vine structure with the marginal distributions was combined and a loss distribution was derived using a simulation approach as discussed in Sect. 3.2. Note, the transformation to pseudo-observations was needed as the marginal distributions of the SPEI are unknown and consequently the copula had to be estimated based on pseudo-samples (see the discussion in Sect. 3.1).

As indicated, the methodology was applied to study drought risk for Corn on the Austrian level for today and in the future. Instead of having a parametric loss distribution, the risk was described in terms of losses and corresponding return periods, e.g., for the 2-, 5-, 10-, 50-, 100-, 250-, 500, and 1000-year events (see Sect. 2.2, Table 4.2). Average production of Corn using the ensemble mean was found to be around 6.2 million tons. The inclusion of climate change opens up an interesting perspective on losses as well as gains. For example, according to the calculations, future Corn yields will increase significantly on average in the future; however, our risk perspective gives an additional more worrying picture, for example, if one compares yields for different return periods. For example, a 500-year event (i.e., low yields which happen approximately every 500 years on average, e.g., a very extreme event) would cause total yields of around 15% lower compared to the average. A comparison with the future yields indicates that while the average is higher in this situation, the 500-year event would decrease by around 20%. In other words, while larger averages in crop yields can be expected in the future, there seems also to be higher fluctuations compared to the current situation, i.e., more and larger extremes, on the system level.

Table 4.2 Corn loss distribution (ensemble mean) on the country level for today and future in million tons (dry matter)

Return period	Current	Future 2050
1000	5.21	6.26
500	5.27	6.46
250	5.33	6.61
100	5.43	6.81
50	5.52	6.97
20	5.66	7.21
10	5.79	7.42
5	5.95	7.67
Average	6.20	8.00

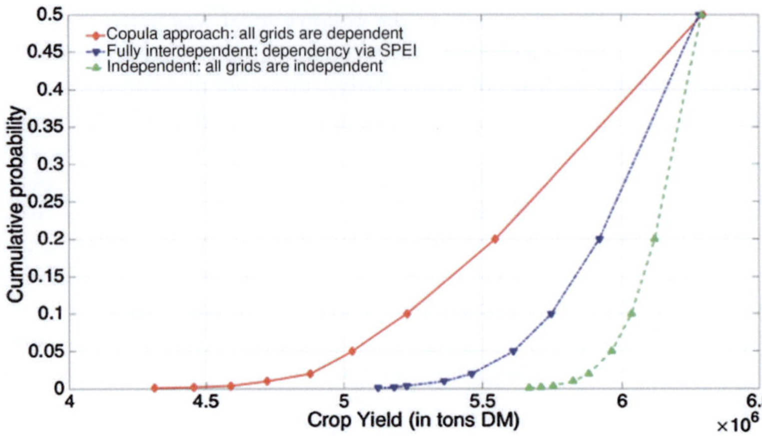

Fig. 4.10 Crop distribution for Corn on the country level assuming different dependencies. *Source* Adapted from Hochrainer-Stigler et al. (2018a)

For determining possible upper and lower bounds of risk due to different assumptions of dependency, calculations for a fully interdependent and fully independent situation were examined (similar to the flood example in the previous section). In more detail (see Fig. 4.10), the fully interdependent approach considered all SimUs in Austria as dependent on each other, regardless of the SPEI value and of possible absence of the long-term influence of drought on crop yields (hence, an upper bound). The fully independent approach considered all SimUs in Austria as independent from each other, regardless of the actual SPEI dependency or crop yield dependency in the region. It therefore represents a lower bound on risk (see Sect. 2.3). The underestimation of risk assuming independency is quite significant, e.g., on average around 20% lower risk compared to the copula approach. However, more importantly, the underestimation of risk gets especially pronounced for extreme events. This ends the discussion of the third application of different copula approaches to upscale loss distributions for individual and systemic risk assessment. As will be further discussed below the estimated loss distributions are key for determining individual and systemic risk through the use of appropriate measures or thresholds. In the next section, some applications based on the results found in this section as well as some additional examples will be presented in the context of risk management.

4.2 Measuring and Managing Individual and Systemic Risk

It was discussed (see Sect. 3.3) that systemic risk can be reduced by decreasing the dependence *between* the individual risks in the system or through the reduction *of* individual risks. For example, in the case of floods changing the dependence could mean to change the river structure or characteristics of it (e.g., soil conditions or water runoff) including modularization (as discussed in Sect. 3.3, e.g., building dykes). The quantification of respective costs is quite complicated and case specific and we rather focus here on some generic options which should demonstrate the main ideas behind such and similar risk management strategies. We now use the results derived in the previous section and apply them for risk management purposes which should illustrate the possibilities of individual and systemic risk reduction by having loss distributions available on different scales. In the first section, we apply the already introduced risk-layer approach (see Sect. 3.4) and demonstrate how it can be used for determining risk reduction and financial risk management options. Here, the loss data from the previous section for the Hungarian regions are taken. Afterwards, we look at the European Solidarity Fund to estimate the ruin probability due to large-scale flood events and consequences on lower scales. Section three presents a fiscal risk approach which is applied on the global level where loss distributions can play an important role as well. The example was also selected as we already have seen in Sect. 2.2 that governments may want to behave risk averse in cases of extremes. Finally, due to some limitations of copula approaches as discussed in Sect. 3.3 at the very beginning, Sect. 4.3 addresses ways forward how loss distributions can be incorporated within agent-based modeling approaches.

4.2.1 Risk Layering on the Regional and Country Level

For this application, we want to take a look at the Hungarian regions and possible risk management measures that could be used to decrease individual and systemic risk. The main ideas are based on Sect. 2.3 and corresponding cited literature there but especially follow Hochrainer-Stigler and Pflug (2012) in regard to risk reduction and insurance modeling. The starting point is the discussion on risk layering from Sect. 3.4 and as for the Hungarian case loss distributions are available it is possible to study risk management options through risk layering. As indicated in Hochrainer-Stigler and Pflug (2012) in principal, two stylized types of functions representing risk management instruments can be thought of: A risk reduction function R and an insurance function I. The consequence of these functions on the loss distribution can be stated in the following way. The risk reduction function will put losses below a given point, say "o" (lower value) to zero and losses that are larger than a given point, say "u" (upper value) cannot be reduced through risk reduction and therefore the full losses will occur (i.e., losses occur as in the case without risk reduction). In between these two points, we assume that the function is linear; however, various other functional forms could be, in principle, considered. For example, risk reduction may be especially efficient near the lower value "o" and efficiency decreases the more the losses are going to the upper value "u" (see, for example, Hochrainer-Stigler et al. 2018d). The risk reduction function $R(x)$ can then be defined as

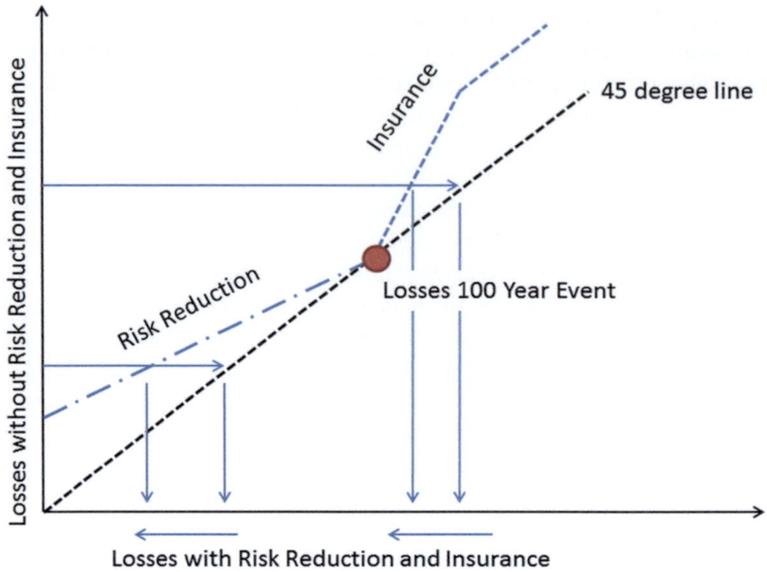

Fig. 4.11 Illustration of a risk reduction and insurance function

$$R(x) = \begin{cases} 0 & \text{if } x \leq o \\ \frac{u(x-o)}{u-o}, & \text{if } o < x < u \\ x & \text{if } x \geq u. \end{cases} \tag{4.2}$$

Similarly, one can define the insurance function "I" with some attachment point a (i.e., the losses where the insurance will start paying claims) and an exit point e (i.e., the maximum losses the insurance will pay claims for) and some pre-defined function that determines how much of the losses between the two points will be actually paid by the insurer (e.g., all losses or only a part of it). We also refer to Fig. 2.10. As in the risk reduction function case, we apply here a common function using a proportionality factor p to determine how much of the losses the insurance would cover. The insurance function $I(x)$ can be defined as

$$I(x) = \begin{cases} x & \text{if } x \leq a \\ a + \frac{b(x-a)}{e-a}, & \text{if } a < x < e \\ x - e + (a+b) & \text{if } x \geq e. \end{cases} \tag{4.3}$$

with $b = (e - a)(1 - p)$. Note, as discussed in Sect. 2.3, one would need to calculate the expected losses of such an contract which could be used to determine the premiums. However, given an utility function, many more options could be considered, including the determination of the attachment point as well as exit point. In practical applications, utility functions are rarely available and rather sensitivity tests are

Table 4.3 Risk measures for different regions in Hungary (in million Euros)

Region	AAL	$VaR_{0.99}$	Expected shortfall
Central Hungary	66	1365	34
Central Transdanubia	18	629	2
Western Transdanubia	69	1582	32
Southern Transdanubia	18	476	5
Northern Hungary	26	861	4
Northern Great Plain	49	1249	17
Southern Great Plain	56	1504	21

performed for calculating premiums for corresponding attachment and exit points. Given the discussion about the risk-layer approach, it can be suggested that for the more frequent events risk reduction is advisable while for larger losses insurance can be seen as more appropriate. We therefore apply a risk-layer approach as discussed in Sect. 3.4 using the introduced risk reduction and insurance function. Figure 4.11 gives an example of both the risk reduction and insurance function on losses.

Given the modeled benefits of risk reduction and insurance on the losses, there is also the question what the costs would be by using such instruments. Regarding the risk reduction part, such analysis is most often done in so-called cost-benefit analysis. One main outcome of such approaches is a cost-benefit ratio that indicates how much losses are decreased by a given investment. According to some literature reviews on cost-benefits of structural mitigation measures, it is sometimes suggested that a ratio of 1 to 4 can be used as an average, e.g., 1 Dollar invested will decrease losses by 4 Dollar (see for a review Mechler 2016). Furthermore, for insurance usually the loadings for different disaster events have to be incorporated to not underestimate the premiums. As indicated, many different techniques can be employed but the basic idea is as given in Sect. 2.3 with the actuarial fair premium to be the average losses for a given insurance scheme. There is also the possibility to determine different packages of risk reduction and insurance schemes and consequences of costs and benefits. For example, if one assumes to decrease risk (as suggested with the risk-layer approach) first with risk reduction up to a 100-year event and afterwards start with insurance till to a given exit level with the rest assumed to be residual risk (see Fig. 4.11), one can calculate the costs and benefits for each layer separately.

Given the loss distributions for the Hungarian regions (see Table 4.1) as calculated in Sect. 4.1.1, we first look at some measures of risk as discussed in Sect. 2.1. For example, we calculate the average annual losses for each region, determine the Value at Risk at the 99% level as well as the expected shortfall given that losses are exceeding the given VaR. We assume that the first loss occurs (e.g., losses greater than zero) after the 25-year event. Table 4.3 shows the results.

This analysis shows that similar risk levels exist both in terms of averages and extremes for Central Hungary and Western Transdanubia, as well as the Northern and Southern Great Plain and the rest having low levels of risks. Let's assume that

Table 4.4 Annual costs for risk reduction and insurance for different regions in Hungary (in million Euros)

Region	RR costs	IR costs	Total costs
Central Hungary	8.14	20.58	28.72
Central Transdanubia	4.17	2.28	6.45
Western Transdanubia	9.25	34.88	44.13
Southern Transdanubia	3.10	3.76	6.86
Northern Hungary	5.54	4.82	10.36
Northern Great Plain	8.00	10.32	18.32
Southern Great Plain	8.81	12.14	20.95

risk reduction can be done up to the 100-year event, and afterwards insurance will be set up till the 250-year event. The corresponding costs can be calculated using the approach above, e.g., assuming a cost-benefit ratio of 4 for risk reduction and for simplicity we assume that all losses are reduced till the 100-year event, e.g., setting "o" equal to "u" and to be the 100-year event. In this extreme case, risk reduction behaves like a jump function with no losses below the 100-year event and full losses afterwards. For the insurance function, we also assume that the proportionality factor is 1 meaning that all losses are financed by the insurance scheme starting at the 100-year event as the attachment point and having the 250-year event as the exit point, and assume that the costs for such a scheme are two times the actuarial fair premium (a rather optimistic case). Also in this case one could interpret, similar to the risk reduction part, that no losses occur for the region as all of the losses would be paid through claim payments. In other words, with such a setting no losses would occur for which the region would be responsible for to pay through their own means up to the 250-year return period. The respective costs for such a setting are given in Table 4.4.

This should exemplify how the individual risk can be measured and managed. Let's assume now that the government wants to assist the two high-risk regions, Central Hungary and Western Transdanubia. Again for simplicity in making the argument, let's assume that the government would decide to be responsible to finance all the losses in case of hazard events. In Sect. 4.1.1, we presented the upscaling approach and we apply it now to these two regions to derive at a loss distribution for both regions. For example, assuming comonotonicity between the two regions a 50-year event on the system level (i.e., the two-node network) would cause losses of around 1600 million Euros while assuming independence between the two regions for all events happening more frequently than a 100-year event, system losses would be around 1100 million Euros. In other words, if the government would make sure that the respective flood event risks in the two regions are independent for more frequent events (here for events that happen on average more often than every 100 years, on average), the risk could be considerably decreased, e.g., by around 500 million. If one assumes that the government would see serious economic repercussions in the two regions for events that happen above the 100-year event (we discuss within the

ABM approach further down below on how such effects could be estimated) eventually 4leading to systemic risks, the corresponding losses under the comonotonicity assumption are found to be around 2.9 billion Euro. Losses could be reduced to 1.7 billion Euro if flood risks are made independent between the two regions for events that happen on average below the 100-year return period. In other words, compared to the full comonotonicity assumption systemic risk could be decreased from the 100- to the 140-year return period. If the government would make sure that risk reduction and insurance is in place as discussed above, systemic risk could be further reduced. Even in the full dependence scenario the systemic risk would realize only at the 210-year event and having no dependency below the 100-year event assumed would decrease systemic risk further to the 310-year event (with the assumption that at 2.9 billion Euro of losses systemic risk may realize). What the actual decision is on how to decrease risk is irrelevant in our context, the important thing is that individual and systemic risk considerations can be incorporated jointly within a coherent decision-making framework.

From an aggregated government perspective, another use of a loss distribution is to assist individual risks on the lower scales through assistance in risk-financing management instruments such as insurance. In Sect. 4.1.3, we calculated drought-related losses on the country scale and one could ask (as it is often done in that regard) if the government may want to subsidize drought insurance products to make it more affordable and therefore prevent ruin of farmers due to such hazard (see for a comprehensive discussion in the case of Austria Hochrainer-Stigler and Hanger-Kopp 2017). One important question is how much such a subsidized insurance system may cost today and in the future which we will address now in more detail. From a risk-layer perspective, this again relates to more extreme events and losses are here defined as crop yields below the long-term average. Austria is quite advanced in terms of insurance products and also has index-based schemes available within the agricultural sector. Note, in high-income countries, insurance products are usually indemnity-based, i.e., claim payments are based on the actual losses. In contrast, index-based insurance products very often use a physical trigger, e.g., rainfall amount over a certain time period, for claim payments to be made. For the later case, the index serves as a proxy for losses and the main benefit is that timely payouts can be achieved with a reduction of administrative costs. One major disadvantage of such index-based products is basis risk, i.e., the situation that the index is not triggered while a loss occurred (Hochrainer-Stigler and Hanger-Kopp 2017). In Austria, especially for extreme events, such index-based insurance products are now introduced and are also subsidized by the government.

We want to answer the question how much such subsidies would cost. We already have seen that drought risk is quite high and may change over time. As currently envisioned in Austria, current and future risk management strategies should support insurance schemes by subsidizing it by 50%. Given that we are assuming that the insurance scheme is providing relief for all yields below the average crop yields, it is possible to calculate losses in terms of deviation from the average (taking only the lower crop yields into account, see also Sect. 2.1). If we assume that the Corn price is around 175 Euros per ton, one can transform the distributions gathered in the

Table 4.5 Government loss distribution due to subsidizing drought insurance (for Corn) for today and future (in million Euros)

Return period	Current	Future 2050
5	25	33
10	42	58
20	55	79
50	69	103
100	78	120
250	88	140
500	94	155
1000	100	175

Source Based on Hochrainer-Stigler et al. (2018a)

previous section into monetary losses (Hochrainer-Stigler et al. 2018a). For example, a 50-year event would cost the government around 69 million Euros, and these costs could increase to around 103 million Euros in 2050. For a more extreme event such as the 500-year event, today costs for the government alone would be around 94 million Euros, again a severe increase in the future to around 155 million Euros in 2050 under the RCP 4.5 scenario has to be expected.

The information gives a detailed picture on drought losses from a system-level perspective based on individual risk on the local level and can also advise future risk management strategies for the government as well as insurance providers in more detail, e.g., backup capital needed for extreme events (Table 4.5). For instance, the actuarial fair premium (AAL) for the government subsidizing insurance is around 13.4 million Euros today and would increase to around 19.9 million Euros in the future. Applying a 250-year event as a reference point for determining necessary backup capital for the government, the fund should be at least capitalized at 88 million Euros today and 140 million Euros in the future. We already discussed that around 20% of the extreme event losses are due to interdependencies during drought events. In case that the government wants to assist insurance providers just for the systemic risk part which is caused due to dependencies other assistance mechanisms could be established as well, e.g., going away from subsidies and taking only the costs of the systemic risk part due to drought events. As also indicated, other options for decreasing dependencies can be established too, for example, irrigation. However, it is not always clear if this is feasible in all regions as there are many complex interactions which would need to be incorporated in such analysis, e.g., decrease in groundwater levels that would put other farmers at risk. Also, some other uncertainties not incorporated here are important in that regard. Modeling wise, the results are based on mean ensemble runs and model ambiguity and price effects were not looked at. As discussed in Pflug et al. (2017) especially in the insurance context model ambiguity gets increasingly important and should be taken into account. Furthermore, to sufficiently compensate farmers to avoid their bankruptcy crop yield prices play a crucial role but are not incorporated in our analysis either. As pointed out in Hochrainer-Stigler et al. (2018a), costs can considerably increase due to these uncertainties. Finally, using such an approach also a detailed analysis of the individual risk levels can be taken as well; however, we refer for the details to the paper of Birghila et al. 2020 and proceed forward with the European Union Solidarity Fund.

4.2.2 An Application to the EU Solidarity Fund

In this section, we look at a special funding mechanism on the European level that was established to assist governments in the coping of losses after natural disaster events. Many similar instruments exist now in the world, including regional pools such as the Caribbean Disaster Insurance Facility (see for a discussion of different instruments Linnerooth-Bayer and Hochrainer-Stigler 2015). The aforementioned instrument on the European scale we are looking at is the so-called European Union Solidarity Fund (EUSF) and we show how the results derived in Sect. 4.1.2 can be used for risk management purposes. Detailed updated information of the fund can be found at the respective websites of the European Commission; however, our presentation here will be mainly based on work done in Hochrainer et al. (2010) and Hochrainer-Stigler et al. (2017c). To set up the stage, we give a short introduction on the history of the fund.

As it is usually the case with disaster risk instruments (Cardenas et al. 2007), also the EUSF was created *after* a large natural disaster event. Indeed, the 2002 flood events that happened mostly in central Europe triggered the political will to institutionalize a form of financial compensation for EU countries that experience large-scale disaster events. The main idea behind the fund was that it should be used as an ex-post loss financing instrument for cases in which a disaster exceeds a government's resources (see also Sect. 4.2.3). Hence, one difference to other such kind of funds is that the focus is on government losses only and not on private sector entities. In more detail, the beneficiary from the fund is only entitled to use the aid to finance operations undertaken by the public authorities to alleviate non-insurable damages. Also different to many other instruments is the fact that it is not set up as a rapid response instrument, also because assistance can only be granted after a series of application and budgetary processes that can take up several months to years (however, the application process was revised to make payments more rapidly, see the analysis in Hochrainer-Stigler et al. 2017c). A threshold approach was adopted for the EUSF to assist which is based on two criteria: either a disaster event happened that caused direct damages above three billion Euros (at 2002 prices) or losses exceed 0.6% of the GNI. Given one of these thresholds is exceeded the amount of aid given for the respective country is calculated as follows: 2.5% of losses are paid for the part of total direct damage below the threshold, and a higher share of 6% for the part of damage exceeding the threshold. The aid is paid out in a single installment after the signing of an implementation agreement with the beneficiary state and should be used within 1 year after the date of receipt.

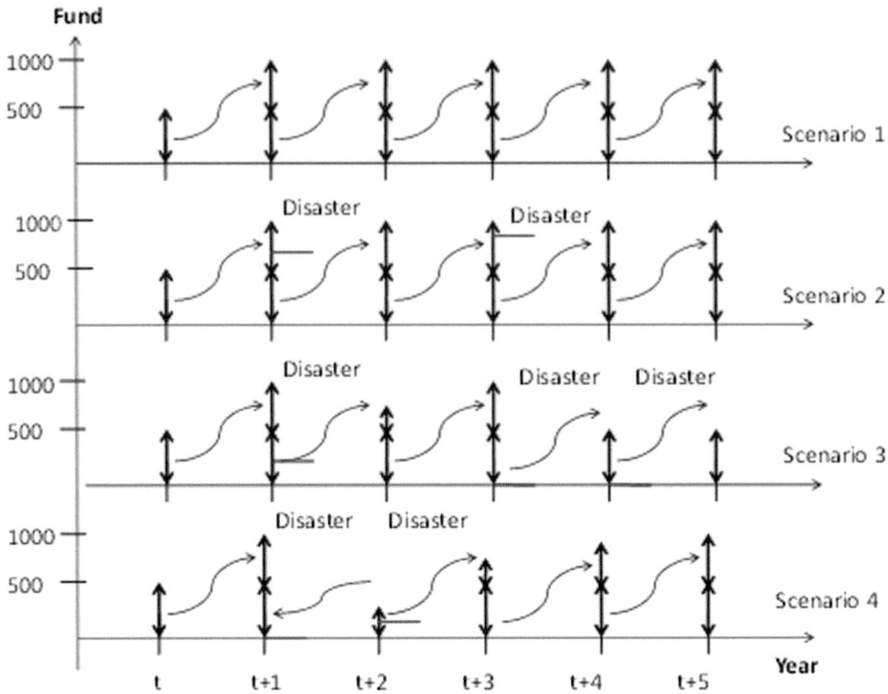

Fig. 4.12 Scenarios of EUSF reserve accumulation processes and disaster events

Since the inception of the EUSF in 2002, it underwent some changes. For example, the EUSF was considered not to be sufficiently responsive, should include drought events, advance payments should be possible, and the EUSF should specify the rules which countries are eligibility under the criterion for regional disasters. To address these and other issues, the commission adopted several reforms in 2014 (for a comprehensive discussion, see Hochrainer-Stigler et al. 2017c). Importantly for our discussion here are the funding rules and capitalization of the fund. Till 2014, the capitalization of the Fund was one billion Euros; this changed and the fund was reduced by half to a ceiling of 500 million Euros (2011 prices) and each year, at least one-quarter of the annual amount shall remain available after October 1. Additionally, a time dimension was included so that funds not allocated in a given year may be used in the following year, but not thereafter. In exceptional cases, the new legislation also allows the use of funds allocated for the following year. This new budget rule is obviously more complicated as the former, especially as it adds a temporal risk-spreading dimension to what was formerly only a spatial risk-spreading instrument. Based on the discussion above, various situations of how the EUSF could perform over time are shown in Fig. 4.12.

Scenario 1 is simply the no-disaster event case, and each year 500 million will be accumulated to the next year only. Hence, the fund would be constantly at the 1 billion level. In Scenario 2, two disasters in year t+1 and t+3 happen; however, they will be financed by the last years accumulated money and therefore no changes in next year budget would happen (as in Scenario 1). In Scenario 3, payments are above 500 million and therefore the next year, the total funding amount will be lower than the 1 billion. Observe, due to repeated events, there is the possibility that the fund will get depleted over time as accumulation is not possible anymore (e.g., years after t+3 in Scenario 3). In Scenario 4, there is the case that the payments are above 1 billion and therefore some money reserved for the following year is used for the current year. Therefore, in year t+2, the actual amount of money is below 500 million, and there is high risk of not being able to pay obligations the EUSF is responsible for. Using a real-world example of the EUSF, we look at 2002. In 2002, approximately 228 million had to be taken from 2003 year which would decrease the fund to 272 million Euro. This would have been enough to finance losses occurred in 2003 and the fund would accumulate to 665 in 2004. Afterwards, till 2009 the fund would be around 1 billion (as only small losses happened in 2008); however, in 2009 only 403 million would have been transferred to the next year, which again would have been enough for 2011 and due to the accumulation also for the high losses in 2012. As already obvious, the amount of funding available gets increasingly uncertain, while the risk of depletion still remains as high as with the other funding scheme, e.g., in the most optimistic case, there is in principle 1.5 billion Euro available which is not increasing robustness very much, especially for very extreme events (Hochrainer-Stigler et al. 2017c).

The EUSF time series of payments to EU member countries is too short for a probabilistic analysis including the assessment if the risk of depletion is negligible. However, given the loss distributions available as done with the copula approach discussed in Sect. 3.3, one is able to perform a stochastic simulation approach for determining the robustness of the fund. As shown in Fig. 4.13, a Monte Carlo approach to estimate the probability of fund depletion based on simulations of disaster events across the EU can be performed (essentially using the inverse transformation method).

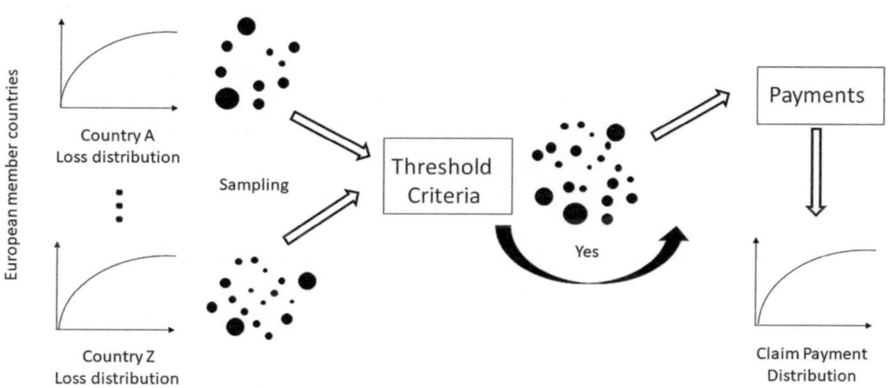

Fig. 4.13 EUSF claim payment calculations. *Source* Based on Hochrainer-Stigler et al. (2017a)

In more detail, one can sample from each country and its respective loss distribution and apply the threshold criteria to choose only those events which are eligible for funding. In the next step, the actual payments can be determined. Through summing up individual annual payments, one can also derive at the fund's annual liability risk, e.g., in the form of a claim distribution. Whether the fund is sufficient to meet the claims can be analyzed in several ways then. Also, a comparison between the pre-reform EUSF can be performed. For example, with the pre-reform EUSF structure, the analysis in Hochrainer-Stigler et al. (2017c) shows that problems occur on average every 28 years, or with an annual probability of depletion of 3.57%. The new EUSF scheme marginally increases the fund's robustness up to the 33-year event, or with an annual probability of depletion of 3.03%. It should be noted that insufficient capital for funding eligible countries may cause problems on the lower levels as well. In regard to dependence, interestingly, it was also found that one could distinguish between different country groups which would be simultaneously affected during especially large-scale flood events (see Jongman et al. 2014). That opens up possibilities to decrease systemic risk if dependencies especially between such country regions are reduced.

As indicated, in case of depletion the EUSF assistance to affected countries cannot be given and therefore countries may fall short of aid in especially extreme cases. This could have important implications on the lower levels, including the government as well as households. However, while for the EUSF the ruin probability could be seen as one measure for determining systemic risk (assuming that in those cases large parts of Europe are under severe stress due to flood events), the consequences on the lower level such as the country scale need to explicitly include some form of coping capacity or resilience (see the discussion in Sect. 2.3). In the next section, we therefore discuss one way forward of determining the fiscal resilience of a country and apply it on the global scale down to the EUSF and country level.

4.2.3 Multi-hazard and Global Risk Analysis

Related to the upscaling issue for loss distributions, a copula approach is also useful to combine loss distributions of different hazards. We want to show one application on the global level which uses loss distributions for different natural hazards on the country level. In more detail, based on multi-hazard loss distributions fiscal stress tests can be performed and global funding needs determined to assist countries in stress situations.

In the most easiest case, one has two annual loss distributions for the same region but different hazards, e.g., earthquake and flooding, that can be assumed to be independent and therefore can be combined through convolution (or setting the p parameter that indicates independence and comonotonicity in Sect. 4.1.1 to the, say 1000-year event) (Fig. 4.14). For other hazards and corresponding losses, there might be some dependencies assumed, for example, hurricane risk and flooding (Woodruff et al. 2013) and here some specific copula types could be used to model this

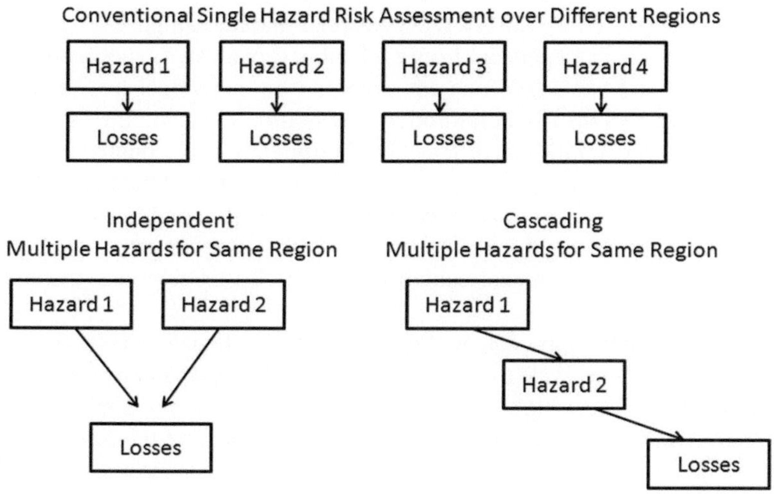

Fig. 4.14 Multi-hazard analysis examples

dependency. In the case of cascading effects, the copula introduced in Sect. 4.1.1 may be of special interest as it can be used to model threshold effects, e.g., a dam brake due to an earthquake event and corresponding larger losses due to downstream flooding (see Fig. 4.14). For a comprehensive risk assessment, a multi-hazard approach can be seen as most appropriate and we want to demonstrate one application in the context of fiscal risks due to natural disaster events through the so-called CatSim (Catastrophe Simulation) approach. The methodology is explained in quite some detail in related publications, see, for example, Hochrainer-Stigler et al. (2015) and we keep the discussion short and focus on the consequences in terms of individual and systemic risks. The discussion is mostly based on analysis done for the Global Assessment Report, especially Hochrainer-Stigler et al. (2017b).

In a nutshell, the so-called CatSim methodology for assessing fiscal risks due to natural disaster events combines estimates of direct (monetary) risks a country is exposed to (e.g., asset losses) with an evaluation of the fiscal resilience of the government, measured through potential resources that are in principal available. The eventual shortfall in financing losses a government is responsible for is measured by the term fiscal or resource gap year event. Regarding fiscal resilience, various financial resources may be available for a country including ex-post options such as diversion from budget or deficit financing as well as proactive instruments such as insurance, reserve funds, or contingent credits. We refer for the estimation of the corresponding parameters to Hochrainer-Stigler et al. (2014b). Regarding direct risk, country loss data from five different hazards, including earthquakes, wind, storm surge, tsunamis, and riverine floods, were available in the form of loss return periods. In more detail, the probable maximum loss (in million USD) for seven different return periods (i.e., the 20-, 50-, 100-, 250-, 500-, 1000-, and 1500-year events, respectively) for each hazard were available on the country level from the Global

Assessment Report (UNDRR 2017). If one assumes that each single hazard represents individual risk and are independent from each other (not always necessarily the case, e.g., earthquake and tsunamis), one can combine the individual loss distributions to a multi-hazard loss distribution through convolution (or using the independence copula). In that way, multi-hazard loss distributions for each country in the world can be derived. For simplicity, it is assumed that around 50% of the total losses have to be financed by the government in case of an event. By combining multi-hazard losses and fiscal resilience, it is then possible to determine the return period where for the first time the government cannot finance its losses anymore, the so-called resource gap year event.

For illustration of the approach, we take a look at two quite different risk situations focusing on the U.S. as well as Madagascar. Absolute risk due to natural hazard events for the U.S. is very high. For example, the 20-year event losses from a multi-hazard loss distribution (earthquakes, wind, storm surge, and flooding) calculated by convoluting all four hazard distributions (via the independence assumption) gives losses to be around an astonishing 121 billion USD. While direct risk seems very large, the U.S. is also very resilient in financial terms, e.g., around 391 billion USD are potentially available to finance losses. As a consequence, the United States do not experience a resource gap at all. In contrast, Madagascar experiences compared to the United States very small losses, e.g., a 20-year event loss from the multi-hazard loss distribution gives losses to be around 0.9 billion USD. While these losses look small, fiscal resilience is very low as well and estimated to be around 168 million USD. Consequently, the risk of not being able to finance losses is much higher for Madagascar with the resource gap year event to be the 5-year event. In other words, Madagascar can expect on average every 5 years that it cannot finance the losses it is responsible for due to natural disaster events while the U.S, despite the high losses, is not at risk of such a situation at all. Note, such analysis has to be done on a continuous basis as the economic situation may change over time and therefore also affects the fiscal resilience of a country. The calculations as discussed above were performed for around 181 countries and Fig. 4.15 shows a corresponding risk map.

According to this analysis, around half of the 180 countries analyzed experienced a resource gap year event below the 100-year event. 56 countries did not experience a resource gap at all, while 32 countries had a resource gap below the 10-year event, which indicates very high fiscal risks. Especially, in the Asian region, the Caribbean, as well as some countries in Latin America and Africa high fiscal risks were found. It should be noted that the losses which are looked at here are direct ones and therefore do not include possible indirect and follow on consequences, which sometimes can drastically increase total losses eventually leading to systemic risk realization (see Sect. 4.3 for a discussion on how to combine the estimates with agent-based modeling approaches). For example, while many developing countries are having a small resource gap size for a 100-year event loss, this does not indicate that indirect consequences will also be small for such events. As already specified above, such effects need to be modeled within a dynamic setting taking all the relevant sectors into account.

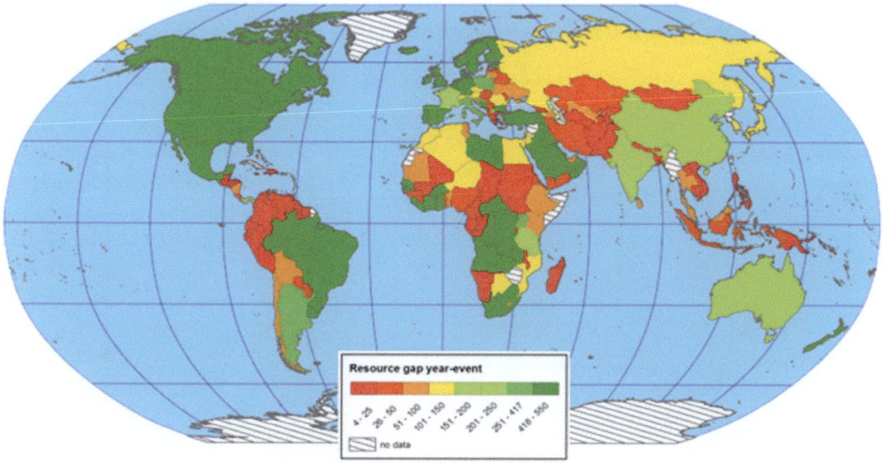

Fig. 4.15 Global analysis of resource gap year events for multi-hazards (EQ, wind, storm surge, tsunamis, and riverine floods). *Source* Based on Hochrainer-Stigler et al. (2017b)

Applying the approach for fiscal resilience as explained above for some of our presented case studies gives some additional indications when systemic risk may emerge. For example, we discussed that the EUSF has a 3% default probability. In case of depletion, Hungary as well as other European countries would need more resources of their own to deal with the events. In the case of Hungary, the resource gap year event would decrease from a 34-year event to a 26-year event (which can be interpreted as individual risk from the European perspective) due to systemic risk on the higher scale. Or in other words, individual risk would occur more frequently, due to large-scale flood events that would deplete the EUSF fund (systemic risk). As indicated above ways to decrease dependencies across regions could be considered to decrease systemic risk levels. Also on the global level, a risk layer may be worthwhile to be considered, e.g., helping the EUSF out in case of large-scale events. As one example, Fig. 4.16 shows annual global funding requirements that would be needed to cover government resource gaps for different risk layers based on the data from Hochrainer-Stigler et al. (2017b).

Similar analysis can also be performed on a country-specific, case-by-case basis (see Sect. 4.2.1). However, even with a risk-layer approach that can provide a basis for planning and action, the complementary roles between public and private investment and the need to address risk for both sectors have to be acknowledged. Attention should be given for exploring the trade-offs implicit in each investment decision including a more detailed analysis of downstream benefits and avoided costs. The suggested approach above is not able to include such interactions explicitly but can be embedded, for example, through agent-based modeling approaches as discussed next.

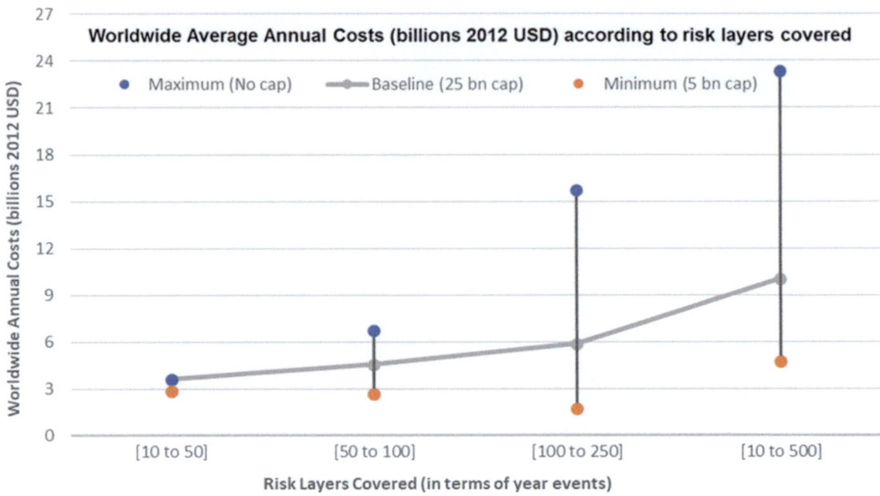

Fig. 4.16 Global funding requirements to assist governments during disasters. *Source* Based on data from Hochrainer-Stigler et al. (2017b)

4.3 Loss Distributions and ABM Modeling

One major limitation of using copulas in the context of loss distributions is that they are not able to model interaction effects of multiple agents and possible emergent and dynamic phenomena directly (see the discussion in Sect. 3.3). One way to circumvent this limitation is to combine the results from upscaling procedures for loss distributions with so-called Agent-Based Modeling (ABM) approaches on higher scales. For example, one can study possible systemic risk realizations due to natural disaster events on different scales now explicitly including the dynamics and interactions of the different kinds of affected agents. In this section, we want to give some indications of how loss distributions and ABM modeling could be combined using some examples for drought and flood events on the country scale. For a thorough discussion about ABM, we refer to the large literature now available on this topic, for example, Epstein (2006) and Heppenstall et al. (2011). Here, the first application will focus on economic consequences and possible systemic risks due to flood events done in a study by Poledna et al. (2018), and the second application will focus on displacement risk in the context of droughts in India done in a study by Naqvi et al. 2020.

The aim of the study by Poledna et al. (2018) was to tackle the question if and how natural disasters can trigger systemic risks on the country level from a macroeconomic perspective. Instead of using risk measures, threshold, or resilience levels to define systemic risk, the cascading events leading to systemic risk were modeled explicitly (rather than through the copula). The ability to simulate the dynamics of how systemic risk actually realizes within a socioeconomic system due to disaster events is one advantage of using an ABM approach and cannot be done with a copula approach. There are also other advantages of ABM approaches as well. From a purely macroeconomic perspective, the classic approaches for determining the indirect effects of disasters are mostly based on econometric modeling approaches as well as input-output-based methods which led to mixed results. For a detailed discussion, we refer to Cavallo et al. (2013) and Rose (2004). ABM approaches have the potential to add to the understanding of such dynamics as they can model individual behavior explicitly; nevertheless, they are quite challenging to be set up on such large scales due to the immense data requirements and the very resource and time-consuming task to calibrate the model. In Poledna et al. (2018), such a highly detailed ABM model for Austria was developed and used for disaster risk and systemic risk assessment due to flood hazards. The basic idea for combining loss distributions including the dependencies of hazard risk through copula approaches with an ABM was quite simple: Within the copula approach, one can determine the probability that many individuals experience losses (e.g., systemic events) and the ABM model determines how the losses to each agent will cascade through the economic system.

In the first step, flood risk on the country level had to be estimated (similar to the previous examples) and afterwards the losses for a given return period had to be distributed to agents within the ABM. As we already discussed in some detail different loss distribution upscaling procedures, we refer to Schinko et al. (2017) on how the loss distribution for Austria was derived (basically using a similar approach as the minimax developed in Timonina et al. 2015b). The loss distribution was used to shock the affected agents through a damage generator approach as depicted in Fig. 4.17. As can be seen, for a given damage event, the distribution of losses to the

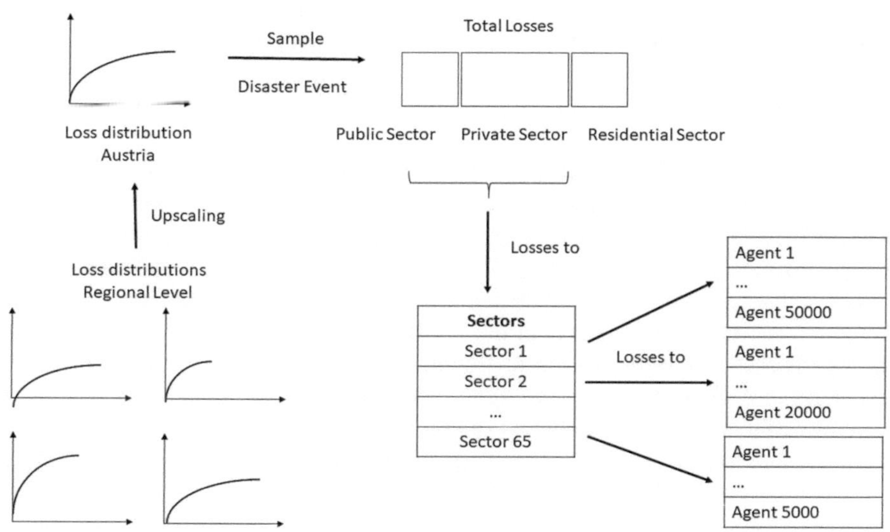

Fig. 4.17 Distribution of total losses to individual sectors

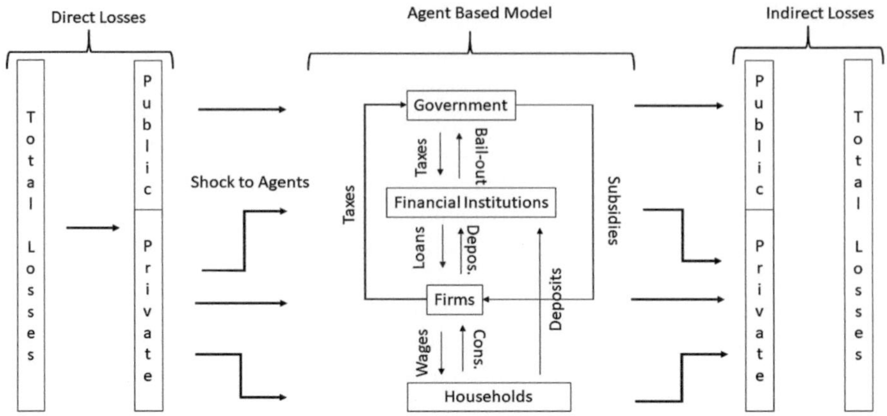

Fig. 4.18 ABM approach. *Source* Adapted based on Poledna et al. (2018)

different sectors and corresponding agents are determined. This was done in a similar fashion as in Lugeri et al. (2010) using an overlay approach (see Sect. 4.1.1). Flood hazard maps for different return periods were employed in that regard which were based on a flood zone system developed in Austria (HORA zoning system) as well as the geospatial distribution of capital and agents according to 65 sectors that are used within the agent-based modeling approach. The ABM used is represented in Fig. 4.18 in a nutshell.

As Fig. 4.18 indicates, the damage generator shocks the individual agents and the ABM subsequently alter their behavior due to the shock which creates higher order effects. These indirect effects can be studied either for different sectors, different dynamics, as well as different time horizons. The detailed ABM model used can be found in Poledna et al. (2018), and here we just look at one important indicative result. Figure 4.19 shows the cumulative changes in GDP growth as a function of the size of direct damages of an event happening in 2014 (the year was used to also have enough real-world data available to evaluate the performance of the ABM in projecting macroeconomic behavior). The focus was on effects up to 3 years after the disaster, including the disaster year. Quite in accordance to previous results found in the literature also here a decrease in GDP growth right after the event can be observed; however, due to increased economic activity (especially for the construction sector due to reconstruction activities after the disaster) some positive overall GDP growth can be observed as well. However, growth is limited due to constraining factors and starts to decline the larger the direct losses will be. We already discussed the issue of resilience within the context of country-level risk. Also here, but now due to the dynamics between the different agents, an inflection point can be identified, that occurs around 2.4% of loss of capital stock. Importantly, the growth stimulus due to disaster can reach a point where resilience is lost and the disaster becomes a systemic event. Note, it is also possible to determine the winners and losers of the disaster event in the short and long run (for a full discussion, we refer to Poledna

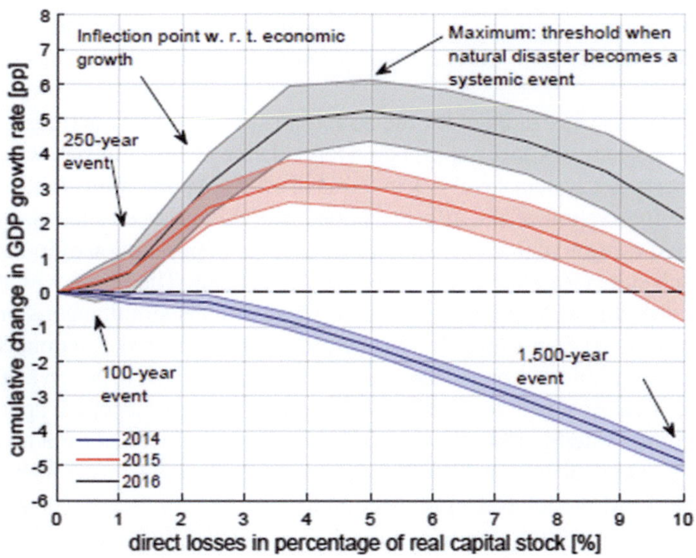

Fig. 4.19 ABM results for GDP change and destroyed capital stock due to flooding. *Source* Poledna et al. (2018)

et al. 2018). Importantly, for the indirect effects, a probability distribution could be established as the input to the ABM is based on the loss distribution which can give the corresponding probability of a given event to happen (however, not done in the study). It should also be noted that the loss distribution and corresponding events are not scenarios but averages of possible scenarios. As discussed before, there are also other possibilities to shock the agents including scenario-based ones. However, these issues are less relevant for our discussion on individual and systemic risk assessment and we move on to show another application of ABMs and loss distribution in the case of drought risk.

While the first example focused on indirect macroeconomic risks due to flood events, our next example looks at displacement risk in the context of droughts. The discussion is based on Naqvi et al. (2020). As in the case of Austria, drought risk was assessed in this study using copulas within a regular vines (R-vines) approach, such as introduced in Sect. 4.1.3, but now on a larger scale, namely, the district level for India which showed large tail dependencies in crop losses in some regions. As in the previous case, the ABM had first to be contextualized, that is, it had to be set up for the specific purpose of analysis. The focus there was on drought-related production shocks, in more detail on shocks in the food system measured through the reduction of crop yields that can cascade through the affected population and more interestingly through the unaffected population too due to interactions. One key feature in the ABM model was the use of income differentials that can result in migration across locations and the model additionally employed an advanced multidimensional network structure that consisted of a household and production layer.

Individual risks were defined for the 36 Indian states, and interactions were modeled with a two-network layer approach. Dependence of drought risk was modeled using copulas to get system-level risk in terms of affected population, while the ABM modeled the interaction of the affected population due to these shocks. Each node was calibrated using actual data to determine the level of production of food and other goods in each state (see Naqvi et al. 2020 for the details). The nodes which produce food are at risk due to a supply-side shock, which means simultaneous production losses in multiple Indian states. In other words, individual risk is connected again with copulas, to mimic a stronger connection during extremes. In Naqvi et al. (2020), two specific scenarios were looked at to test the usefulness of a copula approach and for determining systemic risk. We discuss here simulation results for a no displacement and a displacement scenario and focus on the impact of a drought shock across regions. The ABM model itself is quite complicated and discussed in detail in Naqvi et al. (2020). We just want to note that for the sake of simplicity, the model assumes that all layers adjust completely to demand and supply interactions. Mismatch in markets is corrected through prices, and all output and income is fully exhausted

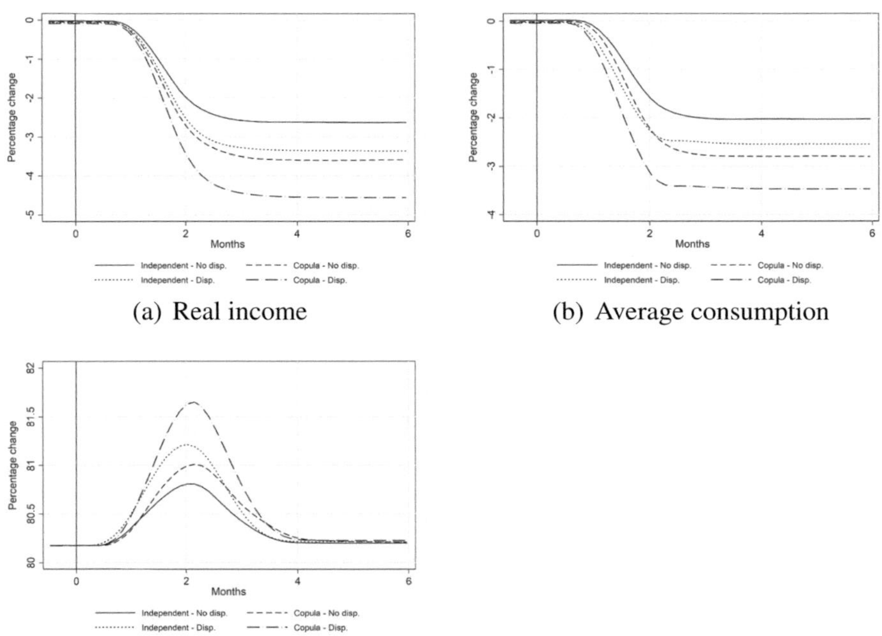

(a) Real income

(b) Average consumption

(c) Share of income on food

Fig. 4.20 Temporal trends from four simulation runs for Indian states. All graphs are shown as percentage changes from baseline no-shock scenario. *Source* Naqvi et al. (2020)

such that there are no leftover stocks. This mimics a general equilibrium framework where prices drive the model outcomes. Within a more real-world setting, also path-dependencies and institutional barriers would play an important role too but were not included. One can interpret a full adjustment model as a best-case scenario model. Regarding results, a shock distribution assuming tail dependence shows on average a larger decline in relative price changes. Since incomes fall, and food becomes more expensive, due to an overall decline in output, the average consumption levels, shown in Fig. 4.20b also fall.

Furthermore, as consumption declines, households allocate more of their income toward buying food to stay above the minimum consumption line. Both figures (Fig. 4.20a, b) show that including or excluding dependencies between states have a much higher impact than with or without displacement assumptions. Figure 4.20c shows the development of this indicator, which rises steeply if one assumes a copula distribution with displacement (this scenario takes the longest to stabilize). These results have the advantage to provide detailed information about the reasons and dynamics of displacement patterns and therefore quite different policy response measures to decrease such kind of risk can be considered. However, in that regard, non-qualitative dimensions come into play which will be the central topic of our final chapter discussed next.

References

Balkovič J, van der Velde M, Schmid E, Skalský R, Khabarov N, Obersteiner M, Stürmer B, Xiong W (2013) Pan-European crop modelling with EPIC: implementation, up-scaling and regional crop yield validation. Agric Syst 120:61–75

Barredo JI, Petrov L, Sagris V, Lavalle C, Genovese E (2005) Towards an integrated scenario approach for spatial planning and natural hazards mitigation. Eur Commun DG-JRC Ispra EUR, 21900

Birghila C, Pflug G, Hochrainer-Stigler S (2020) Risk layering under uncertainty. an application to farmers exposed to drought risk (submitted)

Birkmann J (2006) Measuring vulnerability to promote disaster-resilient societies: conceptual frameworks and definitions. Meas Vulnerability Nat Hazards Towards Disaster Resilient Soc, Chapter 1, United Nations University Press 1:9–54

Borgomeo E, Pflug G, Hall JW, Hochrainer-Stigler S (2015) Assessing water resource system vulnerability to unprecedented hydrological drought using copulas to characterize drought duration and deficit. Water Resour Res 51(11):8927–8948

Cardenas V, Hochrainer S, Mechler R, Pflug G, Linnerooth-Bayer J (2007) Sovereign financial disaster risk management: the case of Mexico. Environ Hazards 7(1):40–53

Cavallo E, Galiani S, Noy I, Pantano J (2013) Catastrophic natural disasters and economic growth. Rev Econ Stat 95(5):1549–1561

Epstein JM (2006) Generative social science: studies in agent-based computational modeling. Princeton University Press

Feranec J, Soukup T, Hazeu G, Jaffrain G (2016) European landscape dynamics: CORINE land cover data. CRC Press

Field CB, Barros V, Stocker TF, Dahe Q (2012) Managing the risks of extreme events and disasters to advance climate change adaptation: special report of the intergovernmental panel on climate change. Cambridge University Press

Gordon W (2011) Calculating catastrophe. World Scientific
Grossi P, Kunreuther H, Windeler D (2005) An introduction to catastrophe models and insurance. In: Catastrophe modeling: a new approach to managing risk. Springer, pp 23–42
Heppenstall AJ, Crooks AT, See LM, Batty M (2011) Agent-based models of geographical systems. Springer Science & Business Media
Hess U, Syroka J (2005) Weather-based insurance in Southern Africa: the case of Malawi. Technical report, The World Bank
Hirabayashi Y, Kanae S, Emori S, Oki T, Kimoto M (2008) Global projections of changing risks of floods and droughts in a changing climate. Hydrol Sci J 53(4):754–772
Hochrainer S, Linnerooth-Bayer J, Mechler R (2010) The European Union solidarity fund. Mitig Adapt Strat Glob Change 15(7):797–810
Hochrainer-Stigler S, Hanger-Kopp S (2017) Subsidized drought insurance in Austria: recent reforms and future challenges. Wirtschaftspolitische Blätter 6(4):599–614
Hochrainer-Stigler S, Pflug G (2012) Risk management against extremes in a changing environment: a risk-layer approach using copulas. Environmetrics 23(8):663–672
Hochrainer-Stigler S, Lugeri N, Radziejewski M (2014a) Up-scaling of impact dependent loss distributions: a hybrid convolution approach for flood risk in Europe. Nat Hazards 70(2):1437–1451
Hochrainer-Stigler S, Mechler R, Pflug G, Williges K (2014b) Funding public adaptation to climate-related disasters. Estimates for a global fund. Glob Environ Change 25:87–96
Hochrainer-Stigler S, Mechler R, Mochizuki J (2015) A risk management tool for tackling country-wide contingent disasters: a case study on Madagascar. Environ Model Softw 72:44–55
Hochrainer-Stigler S, Desai B, Keith W, Maskrey A, Mechler R, Mochizuki J (2017a) Risk-sensitizing future investment needed to achieve the sustainable development goals. Int J Disaster Risk Reduct 24:482–484
Hochrainer-Stigler S, Laurin F, Mechler R (2017b) Fiscal resilience challenged. In: The gar atlas: unveiling global disaster risk. UNDRR, Geneva, pp 47–52
Hochrainer-Stigler S, Linnerooth-Bayer J, Lorant A (2017c) The European Union solidarity fund: an assessment of its recent reforms. Mitig Adapt Strat Glob Change 22(4):547–563
Hochrainer-Stigler S, Balkovič J, Silm K, Timonina-Farkas A (2018a) Large scale extreme risk assessment using copulas: an application to drought events under climate change for Austria. Comput Manag Sci, 1–19
Hochrainer-Stigler S, Linnerooth-Bayer J, Mochizuki J (2018d) Flood proofing low-income houses in India: an application of climate-sensitive probabilistic benefit-cost analysis. Econ Disasters Clim Change, 1–16
Hochrainer-Stigler S, Pflug G, Dieckmann U, Rovenskaya E, Thurner S, Poledna S, Boza G, Linnerooth-Bayer J, Brännström Å (2018e) Integrating systemic risk and risk analysis using copulas. Int J Disaster Risk Sci 9(4):561–567
Jongman B, Kreibich H, Apel H, Barredo J, Bates P, Feyen L, Gericke A, Neal J, Aerts J, Ward P et al (2012) Comparative flood damage model assessment: towards a European approach
Jongman B, Hochrainer-Stigler S, Feyen L, Aerts JC, Mechler R, Botzen WW, Bouwer LM, Pflug G, Rojas R, Ward PJ (2014) Increasing stress on disaster-risk finance due to large floods. Nat Clim Change 4(4):264
Kundzewicz ZW, Lugeri N, Dankers R, Hirabayashi Y, Döll P, Pińskwar I, Dysarz T, Hochrainer S, Matczak P (2010) Assessing river flood risk and adaptation in Europe-review of projections for the future. Mitig Adapt Strat Glob Change 15(7):641–656
Linnerooth-Bayer J, Hochrainer-Stigler S (2015) Financial instruments for disaster risk management and climate change adaptation. Clim Change 133(1):85–100
Lugeri N, Genovese E, Lavalle C, Barredo J, Bindi M, Moriondo M (2007) An assessment of weather-related risks in Europe. Technical report, EUR Scientific and Technical Research Reports, OPOCE

Lugeri N, Kundzewicz ZW, Genovese E, Hochrainer S, Radziejewski M (2010) River flood risk and adaptation in Europe-assessment of the present status. Mitig Adapt Strat Glob Change 15(7):621–639

Mechler R (2016) Reviewing estimates of the economic efficiency of disaster risk management: opportunities and limitations of using risk-based cost-benefit analysis. Nat Hazards 81(3):2121–2147

Mechler R, Bouwer LM (2015) Understanding trends and projections of disaster losses and climate change: is vulnerability the missing link? Clim Change 133(1):23–35

Nakicenovic N, Alcamo J, Grubler A, Riahi K, Roehrl R, Rogner H-H, Victor N (2000) Special report on emissions scenarios (SRES), a special report of Working Group III of the intergovernmental panel on climate change. Cambridge University Press

Naqvi A, Gaupp F, Hochrainer-Stigler S (2020) The risk and consequences of multiple breadbasket failures. Submitted

O'Neill BC, Kriegler E, Riahi K, Ebi KL, Hallegatte S, Carter TR, Mathur R, van Vuuren DP (2014) A new scenario framework for climate change research: the concept of shared socioeconomic pathways. Clim Change 122(3):387–400

Pflug GC, Timonina-Farkas A, Hochrainer-Stigler S (2017) Incorporating model uncertainty into optimal insurance contract design. Insur Math Econ 73:68–74

Poledna S, Hochrainer-Stigler S, Miess MG, Klimek P, Schmelzer S, Sorger J, Shchekinova E, Rovenskaya E, Linnerooth-Bayer J, Dieckmann U et al (2018) When does a disaster become a systemic event? Estimating indirect economic losses from natural disasters. arXiv:1801.09740

Rojas R, Feyen L, Bianchi A, Dosio A (2012) Assessment of future flood hazard in Europe using a large ensemble of bias-corrected regional climate simulations. J Geophys Res Atmos 117(D17):

Rose A (2004) Economic principles, issues, and research priorities in hazard loss estimation. In: Modeling spatial and economic impacts of disasters. Springer, pp 13–36

Rosenzweig C, Elliott J, Deryng D, Ruane AC, Müller C, Arneth A, Boote KJ, Folberth C, Glotter M, Khabarov N et al (2014) Assessing agricultural risks of climate change in the 21st century in a global gridded crop model intercomparison. Proc Natl Acad Sci 111(9):3268–3273

Sadeghi M, Hochrainer-Stigler S, Ghafory-Ashtiany M (2015) Evaluation of earthquake mitigation measures to reduce economic and human losses: a case study to residential property owners in the metropolitan area of Shiraz, Iran. Nat Hazards 78(3):1811–1826

Schinko T, Mechler R, Hochrainer-Stigler S (2017) A methodological framework to operationalize climate risk management: managing sovereign climate-related extreme event risk in Austria. Mitig Adapt Strat Glob Change 22(7):1063–1086

Sklar M (1959) Functions de repartition an dimensions et leurs marges. Publ Inst Statist Univ Paris 8:229–231

Stahl K, Kohn I, Blauhut V, Urquijo J, De Stefano L, Acácio V, Dias S, Stagge JH, Tallaksen LM, Kampragou E et al (2016) Impacts of European drought events: insights from an international database of text-based reports. Nat Hazards Earth Syst Sci 16(3):801–819

Strahler AN (1952) Hypsometric (area-altitude) analysis of erosional topography. Geol Soc Am Bull 63(11):1117–1142

Timonina A, Hochrainer-Stigler S, Pflug G, Jongman B, Rojas R (2015a) Deliverable 3.3: probabilistic risk assessment of high impact events to be used in the case studies of enhance. Deliverable 3.3, ENHANCE Project, Brussels

Timonina A, Hochrainer-Stigler S, Pflug G, Jongman B, Rojas R (2015b) Structured coupling of probability loss distributions: assessing joint flood risk in multiple river basins. Risk Anal 35(11):2102–2119

UNDRR (2017) Global assessment report 2017. Gar atlas: unveiling global disaster risk. United Nations, Geneva

UNISDR (2009) Drought risk reduction framework and actions. United Nations, Geneva

Van Der Knijff J, Younis J, De Roo A (2010) Lisflood: a gis-based distributed model for river basin scale water balance and flood simulation. Int J Geogr Inf Sci 24(2):189–212

Van Vuuren DP, Edmonds J, Kainuma M, Riahi K, Thomson A, Hibbard K, Hurtt GC, Kram T, Krey V, Lamarque J-F et al (2011) The representative concentration pathways: an overview. Clim Change 109(1–2):5

Vicente-Serrano SM, Beguería S, López-Moreno JI (2010) A multiscalar drought index sensitive to global warming: the standardized precipitation evapotranspiration index. J Clim 23(7):1696–1718

Ward PJ, Jongman B, Weiland FS, Bouwman A, van Beek R, Bierkens MF, Ligtvoet W, Winsemius HC (2013) Assessing flood risk at the global scale: model setup, results, and sensitivity. Environ Res Lett 8(4):044019

Winsemius HC, Aerts JC, van Beek LP, Bierkens MF, Bouwman A, Jongman B, Kwadijk JC, Ligtvoet W, Lucas PL, Van Vuuren DP et al (2016) Global drivers of future river flood risk. Nat Clim Change 6(4):381

Woodruff JD, Irish JL, Camargo SJ (2013) Coastal flooding by tropical cyclones and sea-level rise. Nature 504(7478):44

Chapter 5
Conclusion

Our discussion how individual (including extreme) risk and systemic risk can be analyzed in an integrated way focused on quantitative approaches and corresponding techniques. In Chap. 1, we introduced the overall perspective taken in this book by focusing on Sklar's theorem. In Chap. 2, we discussed how to measure, model, and manage individual and extreme risk using loss distributions as the central concept. Afterwards, Chap. 3 presented measurement, modeling, and management approaches for systemic risk and discussed how copulas can be used for the integration of both types of risk. Finally, Chap. 4 presented various applications on different scales and the use of such information for risk management purposes focusing on natural disaster events. In this concluding chapter, we want to discuss less quantitative aspects equally important for the management of individual and systemic risks, namely, human agency as well as governance aspects.

It is a surprising fact that most systemic risk research in the natural-science domain does not, due to various reasons, account for human agency (Page 2015). The ones who did in the past suggested analogies between human systems and biological or ecological ones (see, for example, Haldane and May 2011). Such approaches can be questioned along several lines. Already some time ago, Peckham (2013) questioned the approach to construct banking as an "ecosystem" as appropriate. He argued that by contextualizing financial shocks as a form of "contagion" it creates fear and therefore very much influence policy. The reason why it creates fear is due to the fact that the term contagion reframes financial instability as a form of pathogenicity, thereby re-inscribing socio-cultural and economic relationships as biology. Most importantly, the idea of "contagion" removes human agency and therefore culpability. He therefore concludes that it is problematic to use epidemiological models in financial theory to elucidate the dynamics of turmoil in the markets as such models obscure more than they illuminate, while setting up fears and expectations that are increasingly constraining the parameters of public debate. In a similar vein, Frank et al. (2014) introduces the term "femtorisk" to refer to threats that confront international decision-makers as a result of the actions and interactions of actors that exist beneath the level of formal institutions or outside established governance structures. Such processes confound standard approaches to risk assessment and they argue for

© Springer Nature Singapore Pte Ltd. 2020

S. Hochrainer-Stigler, *Extreme and Systemic Risk Analysis*, Integrated Disaster Risk Management, https://doi.org/10.1007/978-981-15-2689-3_5

creation of new ways of coping with uncertainties that do not depend on precise forecasts of the probability and consequences of future events. They argue especially for adaptive approaches and moving toward an increasingly adaptive risk management framework with focus on solutions with multiple benefits, going away from optimal to a continuous process of adaptation.

In many ways, the measuring, modeling, and managing of systemic risks should be treated fundamentally different in systems that include human agents than in systems that do not (the following discussion is based on Hochrainer-Stigler et al. 2019). The term contingency as pointed out by Ermakoff (2015) for social processes can be seen as a key contribution to the special nature of human agency in the context of systemic risk. He argues that the notion of contingency for social and historical processes denotes a lack of determination: no inherent necessity or master process drives the unfolding events. In other words, events in the past could have happened differently than how they in fact happened; the processes at play in such instances embed an essential causal indeterminacy. Ermakoff specifically looks at the rupture in August 4, 1789, in Versailles and the indecision experienced as a collective state which opened up a range of possibilities including revolution, or in other words systemic risk realization. His contribution is especially important as he was able to embed processes with different types of causation's and corresponding impacts. Partly based on Ermakoff's work, Hochrainer-Stigler et al. (2019) recognizes at least three important and distinct features characterizing systemic risk when human agents are explicitly taken into account: indetermination, indecision, and culpability. For example, free will which creates a fundamental indetermination of human behavior is crucial for systemic risks but cannot be addressed through the tools and methods in the natural-science domain. The self-immolation of Mohamed Bouazizi in Tunisia in December 2010, which started the Arab Spring, is an example of free will—suicide as a form of protest—and what consequences it can have—a systemic event (Pollack et al. 2011).

Hence, especially when it comes to the measuring, modeling, and managing of systemic risks in systems involving human actors, the initiatives of, and interactions among these humans must be considered in detail. Therefore, one of the main differences between the more natural-science related approaches and the human agency perspective is the explicit inclusion of human initiatives and interactions in the risk-producing processes (Hochrainer-Stigler et al. 2019). Similar to Frank et al. (2014) and according to the SREX (2012) one way forward to tackle this challenge is to apply integrative, adaptive, and iterative risk management strategies. As there are very different channels possible, how systemic risk could realize a toolbox-based approach embedded within an iterative process seems one promising way forward. This is due to the fact that a toolbox typically *links* methods, models, and approaches and therefore can highlight the complex nature of systemic risk analyses, especially emphasizing the existence of multiple entry points to the measuring, modeling, and managing aspects of systemic risks. In addition, a toolbox created in this way could provide a new understanding of, and appreciation for, the multiple tools and methods that currently exist. As a consequence, a toolbox approach could be used to contribute to a shift in emphasis from methodology- and technology-focused single means of analysis to an understanding that systemic risk problems are multi-faceted and require a multitude of approaches and methodologies to deliver insights under a broad range of circumstances. Due to the special nature of human agency in regard to systemic risk creation and realization (Hochrainer-Stigler et al. 2019), governance aspects play a pivotal role as well.

Due to the immense literature about governance in general, we restrict our attention to risk governance aspects. Generally speaking, governance refers to actions, processes, traditions, and institutions (formal and informal) by which decisions are reached and implemented (Renn 2017). We already discussed in Chap. 2 that risk usually refers to uncertain, either positive or negative, outcomes (Hochrainer 2006). The combination of the two terms to "risk governance" can be defined (see Florin et al. 2018) as "the totality of actors, rules, conventions, processes, and mechanisms concerned with how relevant risk information is collected, analyzed, and communicated and management decisions are taken." Hence, risk governance deals with the governance of changes/chances and is usually associated with the question of how to enable societies to benefit from change (so-called upside risk, opportunities) while minimizing downside risk (e.g., losses). In contrast (as discussed in Chap. 3), systemic risk is usually seen as a pure downside risk as the realization of systemic risk, by definition, leads to a breakdown—or at least a major dysfunction—of the system as a whole (Kovacevic and Pflug 2015). As systemic risk results from the interactions of individual risks, it cannot be measured by separately quantifying the contributing parts. Therefore, the governance of systemic risk has to look on the interconnected elements and thereby focus the attention on the interdependencies among individual risks (as also suggested in this book). To govern systemic risk Helbing (2013), among others, calls for greater accountability on the part of individual and institutional decision-makers, as well as greater responsibility and awareness of them, for example, through the establishment of the principle of collective responsibility.

However, that may be insufficient if relevant institutions are not set in place. Even if they are in place systemic risks may still realize. For example, while global finance was the best understood and most institutionally developed one in the global governance regimes, these institutions failed to predict or prevent the financial crises in 2007/08 (Goldin and Vogel 2010). However, from a more optimistic point of view, new institutions were developed afterwards to prevent future systemic risks in the financial system, for example, the European Stability Mechanism (ESM) which is an international financial institution by the European Member States to help European area countries in case of severe financial distress. Nevertheless, if no institutional or governance regimes are set in place for a given system which exhibits systemic risks, there are likely no ways forward how to deal and decrease them in the future. Hence, the question is what would be needed to set up institutions to govern systemic risk. While we currently have limited knowledge about how and through what agency systemic risks could emerge today and in the future, they certainly exist (Poledna and Thurner 2016), can increase through specific behavior (Haldane and May 2011), and can represent a major global threat (Centeno et al. 2015).

The situation can be seen similar to the one in the 70s about climate change and how to move forward. Hence, looking at similarities as well as differences between the governance of climate change risks and systemic risks may be fruitful to provide possible steps forward (the discussion is based on Hochrainer-Stigler and Linnerooth-Bayer 2017). One major difference for systemic risk governance compared to climate change governance is in regard to causal attributions. Without causal attribution, accountabilities and responsibilities cannot be established, however, are essentially needed for governing systemic risk. The attribution of climate change was very much related to the analysis of past Greenhouse Gas Emissions (GHG) and commitments and accountabilities could be tackled via GHG projections into the future (Stahl et al. 2016); however, the question how to attribute systemic risks is less clear. Large uncertainties exist in determining the causal effects of systemic risks and usually the complex geography of stakeholders across sectors needs to be explicitly accounted for. This is due to the fact that systemic risk can evolve up to the global level through disruptions at the very small scale or through behavior that is only indirectly linked to the disruption it causes in a specific system. Consequently, the difficulty of attribution bounds the solution space for the reduction of systemic risks as responsibilities and liabilities are unclear; it also hampers the development of a joint vision defining clear targets for managing it.

One practical way forward can be found in the climate risk management community which suggest to bridge the gap between the natural-science and implementation challenges through the use of iterative frameworks (Schinko and Mechler 2017a, Schinko et al. 2017b) and to combine systemic risks with other types of risks so that they can be tackled together (Hochrainer-Stigler et al. 2018b). Especially a closer collaboration between network systems analysis and social sciences to truly enable a system perspective on issues where network dynamics and social processes play an interdependent role seems essential. Furthermore, for the especially deep uncertainties under which systemic risk occurs, an iterative approach would seem to be the most promising way forward for dealing with these uncertainties (see also Jim et al. 2014). As the understanding of systemic risk, from a governance perspective in particular, is still limited (with important exceptions, see the discussion in Florin et al. (2018) and references given there), one suggestion taken from the climate risk community is to use a triple-loop learning process going from reacting to reframing and finally transformation. This is also in line with suggestions made toward an increasingly adaptive risk management framework with a focus on solutions with multiple benefits as discussed before (Frank et al. 2014; Helbing 2013). Based on SREX (2012) and Schinko et al. (2017b), an adapted approach within the context of systemic risks can be suggested (see Hochrainer-Stigler and Linnerooth-Bayer 2017). At the core of the systemic risk governance framework suggested, there are four steps that should ideally be embedded in a comprehensive participatory process (Fig. 5.1).

In the monitoring step, new emerging risks are identified, instruments against systemic risk which are already in place are monitored, changes in perceptions of systemic risks are recognized, and data for calibration and simulation of relevant systems are collected. While this may sound trivial, it is a complex task, given that the dynamic nature and continuous changes in networks and their dynamics, for example, trade networks, can result in large increases in systemic risks within a short time period (Kharrazi et al. 2017). Moreover, in the case of financial systemic risks, the measures implemented may not have the desired effect and may be even counter-productive in the long run. With the data collected, the instruments monitored, and new emerging risks on the agenda, this information is, in the next step, the foundation for quantitative model-based analysis of systemic risks and includes an analysis of ways of measuring, predicting, fixing, and managing them (as these are in effect part of all four steps, they can be organized within the aforementioned toolbox approach, see Fig. 5.1). A quantitative assessment usually has to include some unrealistic assumptions and does not capture possible trade-offs with other objectives real-world decision-makers cannot ignore. Therefore, in the next step, there needs to be an integrated appraisal within a participatory process regarding requisite responsibilities, accountabilities, justice, and solidarity. At this point, it is important to recognize that due to the complexity surrounding systemic risk with respect to warnings of it, predicting it, fixing it, and managing its dimensions, certain special considerations are needed. In this context, one can refer to Page (2015) who argues that, for complex systems and systemic risks, the approaches currently proposed represent a collection of failed attempts.

Nevertheless, one can also emphasize more positively that such approaches do, in fact, represent an ensemble of perspectives, each of which sheds light on some specific

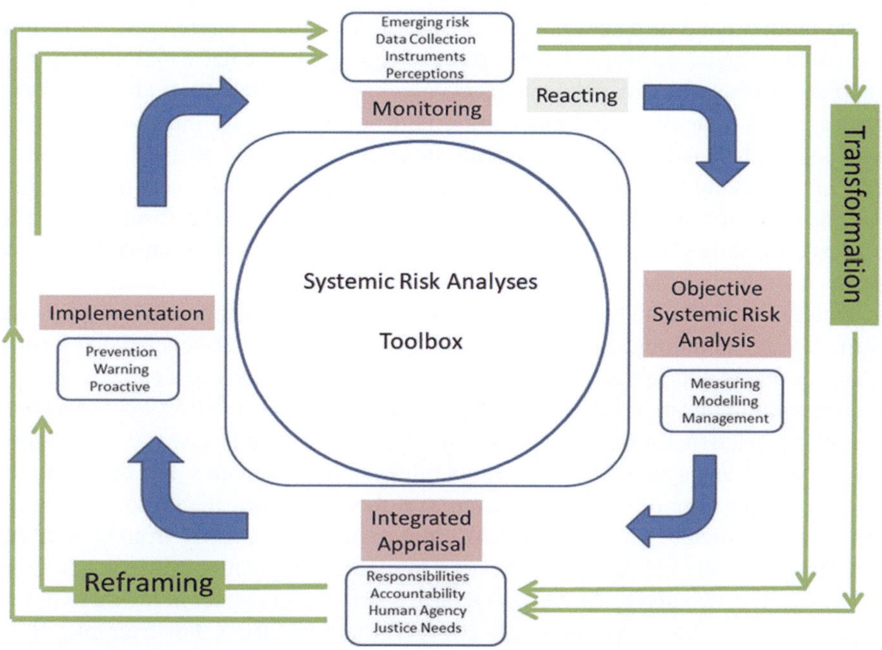

Fig. 5.1 Iterative framework for systemic risk assessment and management embedded in a triple-loop learning process. Iterative framework and learning loop process adapted from SREX (2012); reacting cycle based on Schinko and Mechler (2017a), Schinko et al. (2017b)

aspects of what is itself a complex issue. Consequently, each of the systemic risk measures proposed in the literature, the modeling approaches used, and the different instruments available will have a value in its own right and applying a single measure may inappropriately bias the view on the system. An ensemble perspective is even more important for systemic risk if one acknowledges the fact that the validation of models with past data (e.g., using time series) is difficult to reproduce either through a bottom-up evolutionary system or in a top-down engineered fashion. The idea of an ensemble of systemic risk measures is closely connected to the fact that complex networks usually have the potential to produce multiple classes of outcomes (e.g., equilibria, patterns, complexity, and randomness). An ensemble kind of approach, while not specifically relevant for systemic risks, is appropriate for all problems where large uncertainties exist, for instance, climate change which also uses scenarios and ensembles to deal with epistemic and aleatoric uncertainties (i.e., uncertainties due to lack of knowledge or to the inherent randomness of the process, respectively). In the last fourth step, implementation takes place; this can include an update to the measures already set up or implementation of new prevention, warning, or proactive instruments.

Moving on within Fig. 5.1, if systemic risk cannot be reduced to acceptable levels, if it increases while being monitored, and if new risk potentials emerge, then a more fundamental reconsideration of the current approaches is required: this is depicted as the second loop, namely, the learning loop, which includes a possible reframing of the problem and of respective governance goals. One concrete example of how this could be achieved would be to introduce changes to the network topology, say, imposing taxes within financial markets (Poledna and Thurner 2016). In due course, the proposed systemic risk framework could be extended by integrating entirely new approaches and additional systems into the governance and management of risk, and this could lead to a fundamental transformation of the network structure considered or to systemic risk threats previously identified. Transformation in this context especially should be thought of as a cultural change which consequently also means institutional change (Patterson 2014).

Finally, it should be noted that the term risk and systemic risk include some normative connotations, e.g., risk realization may be seen as beneficial for some while for others not. In that regard, risk as seen as a social construct can be arranged along a continuum from realist to constructivist (Centeno et al. 2015). For the former, a fundamental assumption is the possibility to probabilistically assess the likelihood and impact of any specified risk given its inherent characteristics. For the latter, the existence and nature of risk derives from its political, historical, and social context, i.e., it is constructed. Therefore, risks do not exist independently from society but are created socially in response to the need to regulate populations, interactions, and processes. The two divergent views can have a significant impact in regard to policy implementation (Yazdanpanah et al. 2014; Thompson 2018). For example, Beck (1992) assumes that modernity reflexively relies on increasing complexity to manage the very risks it creates which in turn causes disasters that are often embedded in the very construction of social organizations and institutions. Consequently, iterative approaches are better able to determine conflicts and possible solutions at rather early stages including the right level of abstraction that needs to be taken. Some aspects, such as human agency aspects, may play a less important role in some systemic risk considerations (e.g., in supply chain risks) than in others (e.g., political disruption) which is important to be considered in corresponding governance approaches. The question is related to the optimal complexity to govern systemic risk, i.e., how detailed the approach should be given limited resources. A question that finally can only be answered on a case-by-case basis.

References

Beck U (1992) Risk society: Towards a new modernity, vol 17. Sage

Centeno MA, Nag M, Patterson TS, Shaver A, Windawi AJ (2015) The emergence of global systemic risk. Annu Rev Sociol 41:65–85

Ermakoff I (2015) The structure of contingency. Am J Sociol 121(1):64–125

Florin M-V, Nursimulu A, Trump B, Bejtullahu K, Pfeiffer S, Bresch DN, Asquith M, Linkov I, Merad M, Marshall J et al (2018) Guidelines for the governance of systemic risks. Technical report, ETH Zurich

Frank AB, Collins MG, Levin SA, Lo AW, Ramo J, Dieckmann U, Kremenyuk V, Kryazhimskiy A, Linnerooth-Bayer J, Ramalingam B et al (2014) Dealing with femtorisks in international relations. Proceedings of the national academy of sciences 111(49):17356–17362

Goldin I, Vogel T (2010) Global governance and systemic risk in the 21st century: lessons from the financial crisis. Glob Policy 1(1):4–15

Haldane AG, May RM (2011) Systemic risk in banking ecosystems. Nature 469(7330):351

Helbing D (2013) Globally networked risks and how to respond. Nature 497(7447):51

Hochrainer S (2006) Macroeconomic risk management against natural disasters. German University Press

Hochrainer-Stigler S, Linnerooth-Bayer J (2017) Systemic risk governance (chapter 2.3). In: Global assessment report on disaster risk reduction (GAR), United Nations, Geneva

Hochrainer-Stigler S, Boza G, Colon C, Poledna S, Rovenskaya E, Dieckmann U (2018b) Resilience of systems to individual risk and systemic risk. In: Trump B et al. (eds) IRGC Resource Guide on Resilience, vol 2, EPFL, Lausanne, Switzerland

Hochrainer-Stigler S, Colon C, Boza G, Brännström Å, Linnerooth-Bayer J, Pflug G, Poledna S, Rovenskaya E, Dieckmann U (2019) Measuring, modeling, and managing systemic risk: the missing aspect of human agency. J Risk Res 1–17

Jim H et al (2014) Applied uncertainty analysis for flood risk management. World Scientific

Kharrazi A, Rovenskaya E, Fath BD (2017) Network structure impacts global commodity trade growth and resilience. PloS one 12(2):e0171184

Kovacevic RM, Pflug G (2015) Measuring systemic risk: structural approaches. In: Quantitative financial risk management: theory and practice, pp 1–21

Page SE (2015) What sociologists should know about complexity. Annu Rev Sociol 41:21–41

Patterson O (2014) Making sense of culture. Annu Rev Sociol 40:1–30

Peckham R (2013) Economies of contagion: financial crisis and pandemic. Econ Soc 42(2):226–248

Poledna S, Thurner S (2016) Elimination of systemic risk in financial networks by means of a systemic risk transaction tax. Quant Financ 16(10):1599–1613

Pollack KM, Byman DL, Al-Turk A, Baev P, Doran MS, Elgindy K, Grand SR, Hamid S, Jones BD, Maloney S, et al (2011) The Arab awakening: America and the transformation of the Middle East. Brookings Institution Press

Renn O (2017) Risk governance: coping with uncertainty in a complex world. Routledge

Schinko T, Mechler R (2017) Applying recent insights from climate risk management to operationalize the loss and damage mechanism. Ecol Econ 136:296–298

Schinko T, Mechler R, Hochrainer-Stigler S (2017) A methodological framework to operationalize climate risk management: managing sovereign climate-related extreme event risk in Austria. Mitig Adapt Strat Glob Chang 22(7):1063–1086

SREX (2012) Special report of working groups i and ii of the intergovernmental panel on climate change (IPCC), Geneva, Switzerland

Stahl K, Kohn I, Blauhut V, Urquijo J, De Stefano L, Acácio V, Dias S, Stagge JH, Tallaksen LM, Kampragou E et al (2016) Impacts of European drought events: insights from an international database of text-based reports. Nat Hazards Earth Syst Sci 16(3):801–819

Thompson M (2018) Cultural theory. Routledge

Yazdanpanah M, Hayati D, Hochrainer-Stigler S, Zamani GH (2014) Understanding farmers' intention and behavior regarding water conservation in the middle-east and north Africa: a case study in Iran. J Environ Manag 135:63–72

MIX
Papier aus verantwortungsvollen Quellen
Paper from responsible sources
FSC® C105338

If you have any concerns about our products,
you can contact us on
ProductSafety@springernature.com

In case Publisher is established outside the EU,
the EU authorized representative is:
Springer Nature Customer Service Center GmbH
Europaplatz 3, 69115 Heidelberg, Germany

Printed by Libri Plureos GmbH
in Hamburg, Germany